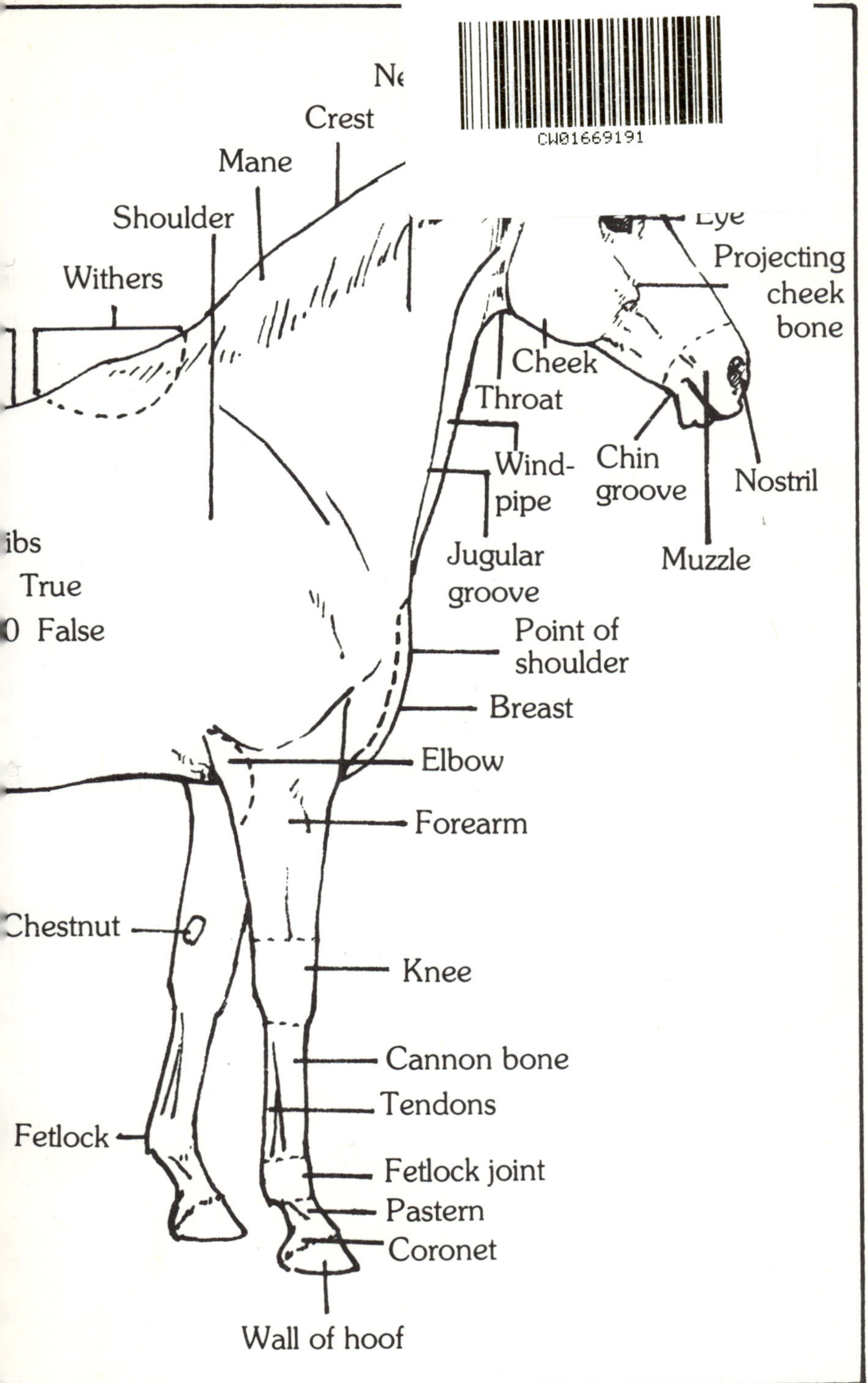
Ne
Crest
Mane
Shoulder
Withers
Eye
Projecting cheek bone
Cheek
Throat
Wind-pipe
Chin groove
Nostril
Muzzle
Jugular groove
ibs
True
0 False
Point of shoulder
Breast
Elbow
Forearm
Chestnut
Knee
Cannon bone
Tendons
Fetlock
Fetlock joint
Pastern
Coronet
Wall of hoof

The British Horse Society
and The Pony Club

THE MANUAL OF HORSEMANSHIP

DELTA BOOKS

Delta Books (Pty) Ltd.
A subsidiary of Donker Holdings (Pty) Ltd.
Hyde Park Corner
Jan Smuts Avenue
Hyde Park
Johannesburg
2196

and
Delta House
35 Hope Street
Cape Town
8001

ISBN 0 908387 29 6

First published 1950
First South African edition 1984

Edited by Marabel Hadfield
Designed by Roger Hyde
Text illustrations by Caroline Bromley Gardner

Produced for The British Horse Society and The Pony Club by
Threshold Books Limited, London.
Filmset in 10 on 11 point Souvenir Light by
Northumberland Press Ltd. Gateshead.
Tyne and Wear
Printed and bound by National Book Printers, Goodwood, Cape.

Contents

Preface

Over thirty years ago a number of distinguished horsemen in England had the idea of producing a book that would form the basis of instruction for Pony Club members. Until then, equestrian training had been based on the military method, and they felt that a different approach was needed to encourage young riders and to improve all-round standards. Thus was born the *Manual of Horsemanship*, this now famous 'blue book' which has sold well over half a million copies, has been reprinted many times, and has helped countless young people to enjoy and improve their riding.

Though each new edition has been carefully checked and amended, there have been no major alterations since the book was first published in 1950. However, over the years new equestrian techniques have evolved, as have methods of instruction, equipment, veterinary medicine, materials and feeding, and it was therefore agreed by The Pony Club Training Committee that the Manual should be brought into line with these changes. Though retaining many of the basic principles and much of the sound common sense of the original, the Committee, with the help of several outside experts, has substantially revised and/or rewritten the text. Other major changes and improvements are a redesigned layout, a completely new set of illustrations, a glossary and a revised index.

We hope that this new version of the Manual will find favour with our many readers and that it will fulfil its primary aim—the pursuit of excellence.

EQUITATION

Equitation
INTRODUCTION

Riding may be enjoyed at many levels. At its simplest it is a means of transport. At its most developed it is one of the supreme partnerships that exist between man and animal.

Few people will deny the beauty and grace of the horse. Most riders come to realise his generosity and gentleness. Those who learn to be true horsemen discover with gratitude his desire to please and serve.

Through man's own lack of thought, ability, knowledge, or kindness, misunderstandings may occur, and bad or ungenerous horses result.

These chapters aim to help all who love horses to become better riders who, in turn, will train better horses.

Good riders must apply themselves, both physically and mentally, to a never-ending task of learning. For The Pony Club member there is a lifetime of pleasure and discovery ahead, as there is for all who wish to learn. Every horse you ride is an individual, and will teach you something new. As your confidence and knowledge increase you will develop resourcefulness, courage and skills within yourself. Above all you will discover the joy of riding.

We ride for our pleasure, but let us do it well. Our horse's and our own enjoyment will then be enhanced.

1

THE RIDER

Mounting and Dismounting
The Position of the Rider in the Saddle
The Position of the Rider in Motion • The Aids

Mounting and Dismounting

MOUNTING

Before mounting see that the girths are secure, the stirrup irons down, the stirrup leathers approximately the correct length and the saddle flaps lying smoothly.

To mount from the near side (the left side of the horse) stand with the left shoulder to the horse's near shoulder. Take the reins and whip into the left hand, with the reins properly separated for riding and of suitable length to prevent the horse from moving. Keep the off-side rein slightly shorter than the near-side rein.

Place the left hand in front of the withers and, with the aid of the right hand, place the left foot in the stirrup. The right hand should turn the stirrup clockwise when guiding the foot. Press the toe down so that it comes under the girth and does not dig into the horse's side. Pivot the body round to face the horse. Place the right hand either on the far side of the saddle at the waist or on the front arch, and spring lightly up, straightening both knees. Swing the right leg over, taking care that it does not brush the horse's quarters, and at the same time position the right hand on the front arch of the saddle. Allow the seat to sink gently without a bump into the lowest part of the saddle. Place the right foot quietly in the stirrup and take the reins into both hands. A rider should be able to mount from either side.

DISMOUNTING

To dismount, remove both feet from the stirrups. Place the reins and whip in the left hand. Place the left hand on the horse's neck and lean forward. Place the right hand on the front arch of the saddle. Swing the right leg back and clear over the horse's back, allowing the body to slip to the ground. Be careful to land on your toes, bending the knees and avoiding the horse's forelegs. Then with your right hand take hold of the reins close up to the bit.

It is correct, and useful, to be able to dismount from either side of the horse.

To Dismount Using the Stirrup

Although this alternative method is given—it may be necessary in the case of a disability—it is not recommended for general use.

To dismount on the near side, take the reins and whip in the left hand and rest them on the horse's neck in front of the withers. Put the right hand on the front arch of the saddle and at the same time slip the right foot out of the stirrup. Pass the right leg over the horse's back and put the right hand on the waist of the saddle. Remove the left foot from the stirrup and allow the body to slip to the ground, as above. It is equally correct to dismount from either side of the horse.

THE STIRRUPS

The Correct Length of Stirrup

The rider can adjust his stirrups to approximately the correct length before mounting by standing facing the saddle, placing the knuckles of the fingers of the right hand on the stirrup bar of the saddle, and adjusting the leathers so that the stirrup iron reaches into the armpit. Before mounting you should stand in front of the horse to check that the stirrups are level.

Once mounted and sitting centrally, allow your legs to hang down in a very relaxed way. You may then adjust your leathers so that the bars of the irons are on or slightly above the ankle bone.

If your leathers are too long, you will lose your balance and therefore your security. You may fall behind the movement in rising trot, lose your stirrups, or grip upwards, your heel coming above your toe.

If your leathers are too short you will appear to be sitting either on top of the saddle rather than into it, or too far back. With a short stirrup you are less likely to be pulled forward or to lose your balance, since you will have the length of your thigh in front of you, but you will also have less use of your lower leg.

As your training advances you will find it easier and more effective to ride with slightly longer leathers. The length will vary according to the standard of the rider and the type of work being carried out.

You should seek the advice of an instructor as to the length at which you should currently be riding.

To Alter the Length of the Stirrup

To alter the right (off-side) stirrup, first take the reins in the left hand. With the right hand take hold of the spare end of the leather, and with the thumb on top of the buckle, steer the tongue of the buckle with the first finger, the other three fingers continuing to hold the spare end of the leather. Disengage the tongue of the buckle and guide it into the required hole. Move the buckle up close to the bar of the saddle by pulling down on the inside of the leather, then replace the end. You should never remove your foot from the stirrup. You should get into the habit of changing the length of your stirrups without looking down.

ASSISTING A RIDER TO MOUNT

Commonly termed 'giving a leg-up'. First see that the girths are secure and the stirrups down.

The rider takes up the reins in his left hand and places it on the horse's withers, with the right hand on the off side of the skirt of the saddle.

A whip or stick, if carried, should be in the left hand.

The rider, standing opposite, squarely facing the saddle, and close to the horse, lifts the left lower leg backwards from the knee.

The assistant stands at the left side of, and facing, the rider, placing his left hand under the rider's left knee and his right hand under the left foot.

On an agreed signal the rider jumps upwards off his right foot, keeping his back straight and shoulders square to the horse's flank.

At the same time the assistant raises the rider's left leg straight upwards, applying the pressure mainly with his left hand and taking care to lift straight up rather than towards the horse.

At this stage the rider must keep his back straight and must not lean forward.

When the rider is able to pass his right leg clear over the horse's back, he turns his body to the front and lowers his seat lightly into the saddle. As the rider turns his body the assistant stops pressing upwards.

The Position of the Rider in the Saddle

It is important for the rider to sit in the correct position, as this will enable him to apply the correct aids, to remain in balance with the horse, and to ride with maximum ease and efficiency.

THE BASIC POSITION

Having mounted, you should sit in the lowest part of the saddle, your hips square with the horse's hips.

You should feel the weight of your body being carried equally on both seat bones. You must be straight *(figure 1)*.

You should look in the direction in which you are going.

Your body, while held upright, should be supple and without tension. It is particularly important that the seat, thighs, and knees should lie relaxed on the saddle as this allows the part of the leg just below the knee to rest against the horse's side.

A rider must at all times remain in balance with the movement of the horse. Suppleness at the hips and flexibility of the spine and shoulders will allow this.

The ball of the foot should rest on the bar of the stirrup iron, exerting just enough pressure to keep the iron in place. Your foot should not tilt to one side or the other.

The line of the foot from heel to toe should point directly forwards. The ankle must remain supple, and the heel should be slightly lower than the toe.

Seen from above and from the side a straight line would pass from the rider's elbow, to his hand, and along the rein to the horse's mouth *(figures 2a and 2b)*.

(a) (b)

1(a) *Seen from behind, a line drawn down the rider's spine would pass through the horse's spine.* **(b)** *Seen from the side, there should be a line that is at right angles to the ground and that passes through the rider's ear, shoulder, hip and back of heel.*

It is vitally important that you should be able to move your hands independently of your body. Your hands must move in harmony with the horse's mouth. When the horse moves his head and neck your hands must be with that movement.

This hand movement is made possible by the suppleness and mobility of your shoulders and elbows. Wrists must remain supple, but they should not bend.

When you take up the reins you should feel some weight in your hands. Ideally, you should have the same feel in your hands at all times and at all paces. This is known as the 'contact'.

It is important that a horse should accept this contact happily and feel no pain or discomfort from it. He will do so if the rider is able to 'go with' all the movements of the horse's head and neck. (This is possible only when a rider has attained an independent seat.)

The contact must be neither too light nor too heavy, but must give a horse a comfortably secure and sure feeling.

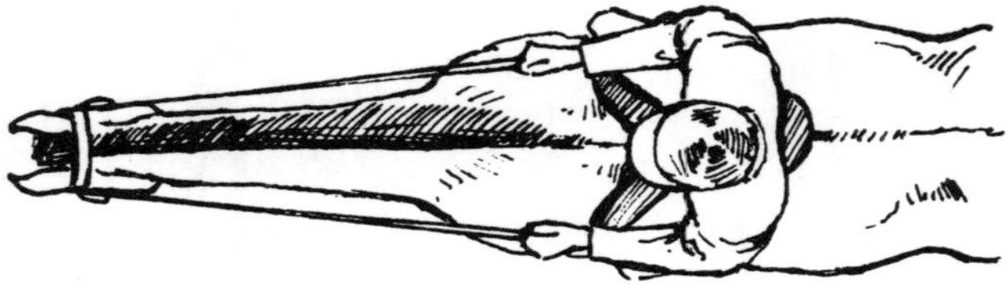

2(a) *Seen from above, there must be a line that passes from the elbow to the hand, along the rein, to the horse's mouth.*

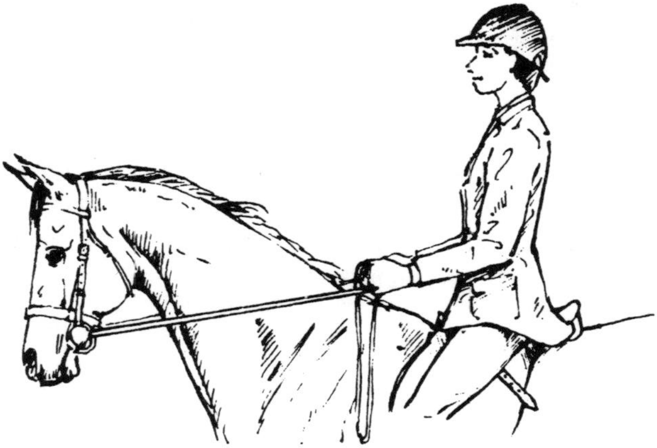

2(b) *Seen from the side, there must be a line that passes from the elbow, along the rein to the horse's mouth.*

A rider must also make alterations to the height of his hands as the horse moves his head up or down, thus maintaining the line—'elbow, hand, horse's mouth.' Both hands and legs should always blend with the movements of the horse unless applying an aid.

Your hands should be carried with the thumbs uppermost and the backs of the hands facing directly outwards. You should hold your wrists so that there is a straight line down the forearm and back of the hand.

You should not allow your wrists to stiffen or to become tense, as this will only tend to make your elbows and shoulders tense and rigid.

THE REINS

Holding the Single Rein

The rein must pass directly from the bit, between the little and third fingers, then across the palm and over the index finger, the thumb being placed on top. It is important that your third finger holds the edges of the rein in the joints nearest the palm and that your fingers are closed securely but without tension. It is the holding of the rein in the third finger that stops it slipping through your hands *(figure 3)*.

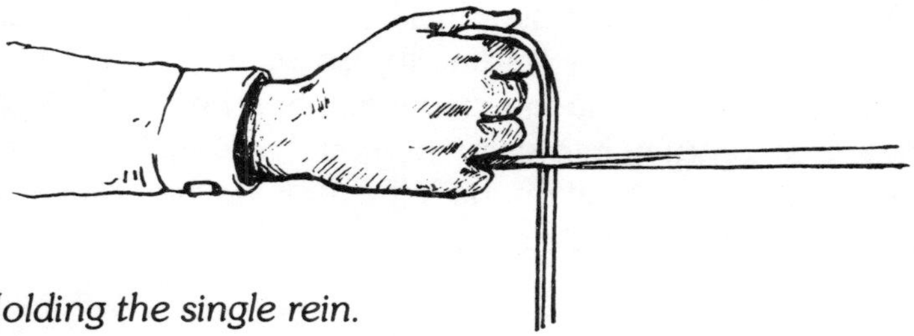

3 *Holding the single rein.*

Holding Two Reins (Double Bridle, etc.)

Hold as for the single rein, except that the little finger of each hand should divide the reins. The bridoon rein is usually held on the outside.

To hold all the reins in one hand, pass the reins from your right hand into your left hand so that the second finger divides them, with the slack ends passing over the index finger and secured by the thumb.

The Position of the Rider in Motion

IN WALK

The position of the body does not alter except that it moves slightly at the hips and waist in rhythm with the natural movement of the horse.

The elbow and shoulder joints move considerably, allowing the hands to go with the natural movement of the horse's head and neck.

THE RISING TROT

When correctly performed, the rising trot is an easy movement for both horse and rider.

The trot is an alternate movement of a horse's two diagonal pairs of legs. The near hind and the off fore are the 'right diagonal', and the off hind and near fore are the 'left diagonal'.

A rider is said to be riding on the 'left diagonal' when the off hind and near forelegs come to the ground, and on the 'right diagonal' when his seat returns to the saddle as the near hind and off forelegs come to the ground *(figure 4)*.

The rider changes the diagonal by sitting down in the saddle for an extra beat before starting to rise again.

The rider should change diagonal when he changes direction, and at frequent intervals when out hacking. It is generally considered correct to sit on the left diagonal when on the right rein, and on the right diagonal when on the left rein.

4 *The rider sitting on the 'right diagonal' in rising trot.*

In rising trot your upper body should be inclined slightly forward from the hips so that you remain in balance with the horse's movements. While actually rising, your shoulders should not be further forward than your knees.

Your body should be raised by the movement of the horse, your seat returning quietly to the saddle without any loss of balance.

Be careful not to collapse at the waist, causing a rounding of your back. Do not allow your weight to fall back into the saddle, thus coming behind the movement.

Hip and knee joints must remain supple and mobile while opening and closing, to accommodate the raising and lowering movement.

The weight on the stirrup irons should not vary.

The contact of the lower legs should not vary.

Elbow and shoulder joints should be supple and mobile, allowing the hand to maintain the correct position. As you rise, the angle of your elbow joint must open, closing again as you return to the saddle.

Your hands should maintain the same contact at all times.

THE SITTING TROT

You must remain completely upright, your legs and hands maintaining perfect contact.

Hips and back should remain supple, absorbing the movement of the horse.

Sitting trot is used on the more trained horse to allow the rider to remain close to the saddle at all times. It should only be used when the horse's back is sufficiently muscled to carry a rider in this way. If sitting trot is used before the horse is ready, he will 'hollow' his back and raise his head, thereby losing suppleness, relaxation, and effective use of his hocks.

THE CANTER

At this pace, suppleness of the hips is most important. Your upper body should move in rhythm with the three beats of the horse's stride (see page 27).

The seat should remain close to the saddle for all three beats, the back and hips allowing this through their suppleness. The suppleness and mobility of the shoulder and elbow joints are also very important, as there is considerable movement of the horse's head and neck.

If the rider stiffens his back he will bump in the saddle, which is most uncomfortable for both horse and rider.

THE GALLOP

Until a horse is going forward freely and has sufficient impulsion the rider should maintain the canter position.

When the gallop is established, the weight of the rider's body should be taken forward on to the knees and stirrups. The weight of the seat should be just off the saddle.

The reins may require shortening to allow the line 'elbow-hand-horse's mouth' to be maintained.

It is easier for the horse if this forward position is adopted, as your weight will be poised over the centre of gravity. To maintain the position you will need a slightly shortened stirrup leather. You must be fit. You must not lose your balance and bump on the horse's back *(figure 5)*.

5(a) *The gallop: correctly shortened stirrup length.*

5(b) *The gallop: the rider losing balance.*

TRANSITIONS (Changes of Pace or Speed)

During transitions it is important that your body should remain in perfect balance with the horse's movement.

In transitions up and down you must remain quiet and supple, neither anticipating nor getting left behind the increasing or decreasing pace. You must resist the temptation to influence the horse by throwing your body about and by looking down or leaning in. This particularly applies to transitions from trot to canter.

CIRCLES AND CHANGES OF DIRECTION

You should allow your weight to remain equally on both seat bones and to stay correctly in the centre of the saddle, sliding neither to the inside nor the outside.

Hips must remain parallel to the horse's hips and shoulders parallel to the horse's shoulders. You must look in the direction in which you are going.

The 'inside' and 'outside' The 'inside' is the side of the horse that is on the inside of the movement, i.e. when on a left circle the inside is the left (near) side of the horse.

The 'outside' is the side of the horse that is on the outside of the movement, i.e. when riding on a left circle the outside is the right (off) side of the horse.

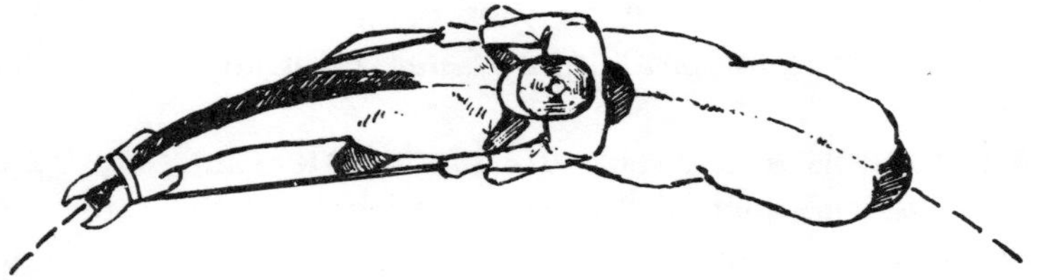

6 *Travelling on a left circle, showing the correct bend.*

The 'bend' of the horse When moving on a curve, a circle, or round a corner, the horse must bend his body uniformly from poll to tail to comply with the curve he is travelling on. This is termed the 'bend' *(figure 6)*.

The bend is considered incorrect when the horse bends his body away from the direction in which he is travelling.

The Aids

Aids are the 'language' used by the rider to communicate with the horse. The aim should be to give quick, clear aids at all

times, and to receive immediate response from the horse. There are two types of aid:
Natural aids: The rider's legs (lower), hands (fingers), the influence of the body and seat, and the voice.
Artificial aids: Whips and spurs.

NATURAL AIDS AND THEIR APPLICATION

The Legs

Relaxed thighs and supple hip and knee joints allow the legs just below the knee to rest in equal contact against the horse. The contact of the leg must be sure, definite and unvarying, except when you are giving aids.

The inside leg The inside leg asks for impulsion with a quick, light, inward nudge applied at any moment needed.

A slightly increased pressure inwards encourages the horse to bend correctly.

The outside leg The outside leg influences the quarters in two ways:

- To ask for canter it brushes back and gives a definite nudge behind the girth, immediately returning to its normal position.
- It controls and directs the hindquarters laterally by an increased contact of the lower leg behind the girth. The pressure of the leg continues until the correction or movement sideways has been completed.

The Hands

Supple shoulder and elbow joints allow the hands to follow the movement of the horse's head and to maintain an equal contact on the reins at all times.

The inside hand The fingers of the inside hand ask for direction with a quick take-and-give.
The outside hand The fingers of the outside hand control and regulate the speed and pace with a quick take-and-give, take-and-give. The outside hand also allows and controls the bend of the horse.

When giving finger aids the arms must still continue to follow the movement of the horse's head.

Hand and Leg Aids Summarised

To walk The rider asks for walk by creating impulsion with his legs and by following the movement of the horse's head and neck with his hands. He controls the speed with his outside hand.

To trot As for to walk.

To canter (direction left) The rider makes sure that the horse is going with impulsion, then gives a small aid with the left rein and moves the right leg back to give the canter aid. (For direction right, give the aids with the right rein and the left leg.)

To decrease speed or pace The rider asks with a quick take-and-give, take-and-give, of the outside hand.

Influence of the Seat

It is important for young riders to learn how to sit naturally and softly, and to blend their body movements harmoniously with those of the horse. On more trained and muscled horses the seat aid involving the back, correctly used, can produce increased impulsion. It should be used with discretion. The rider must always remain in balance and be aware of the influence of his seat and weight.

The Voice

The voice should be used to praise, soothe, or scold the horse. In the early stages of training it should be used in conjunction with the other aids, to clarify them, saying for example, 'walk on', 'trot', 'canter', 'steady' or 'halt'. Later, when the horse understands the leg and hand aids, the voice is used to praise, and it can be associated with a pat.

These aids, logically applied, may be used without any alteration for all paces and movements at all standards of classical riding.

ARTIFICIAL AIDS AND THEIR APPLICATION

The Whip

When a horse disobeys the rider's leg aids, the aid should be repeated and at the same moment the whip used just behind the leg that gave the aid, i.e. behind the rider's inside leg for impulsion; behind the rider's outside leg for control of the quarters or canter strike-off.

There must be no delay in using the whip; you must therefore become very adept at handling it. The whip should normally be carried in the inside hand, as the inside leg is the one most frequently disobeyed. You must therefore change the whip to the correct hand when you change direction. When the horse responds to the whip by moving forward vigorously it is important that you allow the movement, and praise him. In this way he will learn from the correction.

A horse must never be hit in temper.

To change a long whip from one hand to the other in the simplest, most efficient way, the whip should be passed over the horse's withers, the point making an arc directly above the rider's hands. For example, to change the whip from the right hand to the left hand:

Take both reins into the right hand.

Pass the left hand over to take the whip, grasping it so that the thumb and index finger come below the right hand.

Release the whip from the right hand, return the left hand to the normal position and re-take the left rein.

Types of Whip

The standard type of whip Approximately 0.74m (2½ft) long, this is suitable for general use and for jumping. It should be held with the knob uppermost. The hand holding the whip must be removed from the rein when the whip is applied.

The dressage, long schooling or polo whip Approximately 0.91m (3ft) long, this is used in experienced hands for training. With experts it may be used while keeping both hands on the reins. It is held with the knob uppermost, and the hand close to the knob. It should not be used when jumping.

The hunting whip This should never be carried without its thong and lash. The thong is usually of plaited leather, and the lash a small piece of silk or whipcord attached at its end.

The hunting whip should be held at the point of balance, with the hook below the hand, the thong lying through the hand and hanging downwards.

The hunting whip has several uses:

- To reinforce the leg aids.
- To help the rider when opening, holding open, and shutting gates.

- To crack, or to make tapping noises against the saddle when cub-hunting.
- To keep hounds off the heels of the horse.

The stick (or cane) This is used principally for showing. The plain stick or cane is held at the point of balance.

The Spurs

There are many types of spur. Those with rowels should never be used. Blunt spurs may have short or long necks; some have rounded ends.

Use of spurs Spurs should only be worn and used by riders who have both an independent seat and complete control of their legs.

They are used by experienced riders to reinforce the leg aid and should only be used when required.

Spurs should be used without force, applying the inside against the horse's side. To do this the rider turns his toe fractionally outwards. If the toe is turned outwards too far the back of the spur will touch the horse's side. This is incorrect.

Fitting Spurs are usually slightly curved and should always be worn pointing downwards, the longer side being worn on the outside of the boot. The spur straps should be long enough to allow the spur to lie along the seam (counter) of the boot and should be the same colour as the boot. The buckle of the spur strap should be fitted as close to the outside of the spur as possible *(figure 7)*.

7(a) *A spur correctly fitted.*

7(b) *A spur incorrectly fitted.*

2

THE HORSE

The Correct Way of Going
The Basic Paces

The Correct Way of Going

When the horse is going correctly he should present a picture of perfect harmony and grace. He should never look unnatural, and should accomplish all that is asked of him without apparent effort.

He should be a pleasure to ride.

8 *The horse should be a pleasure to ride.*

ACCEPTANCE OF THE BIT

By accepting the bit, the horse accepts the contact of the rider's hands on the reins. His whole body responds to this influence from his quarters through his back to the bit.

There must be no resistance in the mouth. The horse must respond to the rider's hand (finger) aids willingly and without hesitation.

STRAIGHTNESS

When moving directly forward the horse should go as if on railway tracks with the hind foot on one side directly following the fore-foot on the same side.

He should be straight from the tip of his nose to his tail.

When moving on a curve he should bend uniformly from poll to tail. The footprint of the hind foot must follow the footprint of the forefoot.

The smaller the circle, the more acute the curve of the horse's body. It is important that he does not bend his neck more than his body.

Straightness is maintained by the horse going forward in response to the contact of the rider's legs.

BALANCE, IMPULSION, RHYTHM AND TEMPO

Balance

A horse is said to be in balance when his own weight and that of the rider are distributed in such a way as to allow him to use himself with maximum ease and efficiency.

A young horse when at liberty learns to balance himself. He must, therefore, learn to re-balance himself when he is carrying a rider.

This is achieved by the progressive training of the horse, resulting in greater development of the correct muscles and increased use and engagement of his hind legs.

A horse with too much weight on his forehand is considered to be out of balance.

By the suppleness and mobility of his back allowing the engagement of his hind legs, his weight is moved off his forehand.

Untrained horses frequently lose balance when changing direction or pace.

Too much or too little impulsion may result in the horse losing balance.

A horse is in balance when he carries himself with ease.

To improve a horse's balance, you must be ready to make alterations to the impulsion and to the speed (within the pace), thus encouraging the horse to make his own adjustments to correct his own balance.

An excellent method of getting a horse in balance is by riding up and down hills.

The horse should be allowed to find his own balance and the rider should in no way attempt to raise, lower, or position the horse's head with the hands. The way a horse carries his head is merely an indication of the way he is using his body. When he carries himself correctly, the head and neck will automatically become part of the whole pleasing picture.

Impulsion

Impulsion is energy: energy asked for by the rider, and supplied by the horse. Impulsion can be produced only when everything else is correct.

A horse is going with the correct amount of impulsion when he is willing to go forward actively and vigorously: his back is supple, his hocks are engaged, and his steps show elasticity. He must be without tension, and both mentally and physically relaxed.

You must be careful not to confuse impulsion with speed. The rider increases impulsion mainly with the inside leg (see Aids).

Speed is controlled mainly by your outside hand (see Aids).

Rhythm

Rhythm is the regularity and evenness of the hoof-beats.
The horse must maintain his rhythm at all times.
This is achieved by the horse maintaining his balance, and by the rider being aware of the slightest alteration and making adjustments to either the speed or the impulsion.

Tempo

Tempo is the speed of the rhythm.
The tempo of each pace should remain the same.

The tempo is maintained when the horse stays in balance. This is achieved by the rider's careful adjustment of speed and impulsion whenever necessary.

THE OUTLINE

The correct outline, or profile, is achieved when, in motion, the horse is accepting the bit, and his movements are straight, rhythmic and balanced. Seen from the side the outline should show:

Upper line of the body:

- The poll as the highest point.
- The front of the face never behind the vertical.
- The neck curving gently from withers to poll.
- The back and loins supple and relaxed.
- The tail soft and swinging.

Lower line of the body:

- The jaw relaxed and quiet, without tension.
- The muscle underneath the neck soft and relaxed.
- The shoulders looking light, unweighted and moving freely.
- The joints of the forelegs, elbows, knees and fetlocks moving equally.
- The hind legs looking springy and never seeming over-burdened.
- The joints of the hind legs, stifle, hocks and fetlocks all moving equally.
- The hind legs being picked up energetically as the horse's body passes over them and not left to dwell on the ground behind him.

The Basic Paces

At all times the horse should accept the bit and the contact with the rider's legs and hands.

THE WALK (MEDIUM)

Medium walk is the pace between collected and extended walks.

There are four beats to a stride, so the walk is termed 'four-time'. These beats should be even and regular so that the rider can count 1-2-3-4-1-2-3-4-. If there are any irregularities in the beat it is a serious fault.

The sequence of foot-falls is: near hind, near fore, off hind, off fore *(figure 9)*. The horse always has at least two feet on the ground at the same time.

The walk should look and feel calm, active and purposeful. The horse's hind foot should pass over the print left by the fore-foot on the same side. This is called 'overtracking'.

There is considerable movement of the horse's head which must in no way be restricted by the rider.

Should you choose to offer the horse the relaxation of his head and neck by giving the rein away, the horse should stretch his head and neck as if 'looking for the bit'.

Aids to the Walk

From halt the rider asks for the walk by creating impulsion with the legs and by following the movement of the horse's head and neck with his hands.

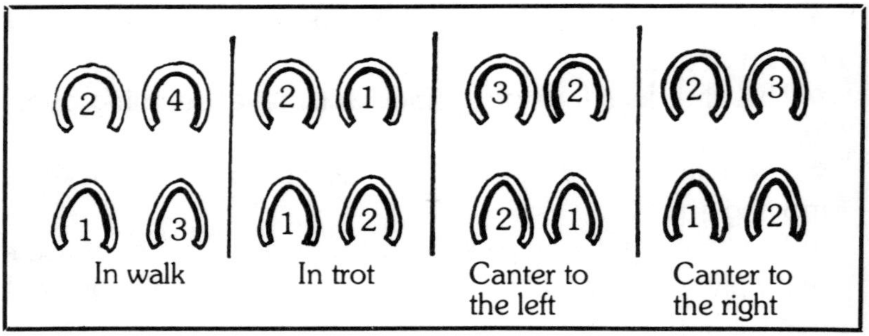

9 *The sequence of foot-falls.*

THE TROT (WORKING)

Working trot is the pace between collected and medium trots.

The previous pace must be of good quality.

There are two beats to a stride, so the trot is termed 'two-time'. The horse springs from one diagonal pair of legs to the other with a moment of suspension between each step. The steps must be regular and even.

The sequence of foot-falls is: near hind and off fore together, then off hind and near fore together *(figure 9)*.

If the trot tempo is fast and hurried it is termed 'running'.

If the trot tempo is slow and the moment of suspension too long, it is termed 'swimming' or 'elevated'.

Both are incorrect.

Aids to the Trot

From halt or walk the rider asks for the trot by creating impulsion with the legs (a quick inwards nudge) the hands going with the movement of the head, and in no way restricting the change of pace. The pace is maintained chiefly by the use of the rider's inside leg.

THE CANTER (WORKING)

The working canter is the pace between collected and medium canters.

There are three beats to the canter stride, so the canter is termed 'three-time'. When listening to the hoof-beats you can hear 1-2-3- and afterwards a moment of silence when all the hooves are off the ground at the same time. This is called the moment of suspension.

In the canter the horse should appear light on his feet, and be balanced and rhythmic. There should be considerable movement of the horse's head and neck.

Sequence of foot-falls When the horse is leading with the off leg, the sequence is: near hind; off hind and near fore together; off fore. When leading with the near leg, the sequence is: off hind; near hind and off fore together; near fore *(figure 9)*.

Aids to the Canter

Before asking for canter, check that the preceding pace is of good quality.

The horse must be accepting the bit and going forward in balance and with impulsion. You should indicate with your inside hand the direction of the canter (with a quick take-and-give), sit for a few strides, move the outside leg back behind the girth, and give a definite nudge to the horse's side. As the horse strikes off into canter you will feel the alteration of pace and you must be particularly careful to remain supple, relaxed and in balance. Your hands must go with the considerable movement of the horse's head; the impulsion is maintained chiefly by your inside leg.

You must on no account look down to see which leg is leading; you must learn to feel which shoulder of the horse is slightly in advance of the other and which hind leg comes to the ground first.

Terms Used When at Canter

A horse should always canter 'united'. A horse is said to be cantering 'true' or 'united' when the leading foreleg and leading hind leg appear to be on the same side. He is said to be cantering 'disunited' when the leading hind leg appears to be on the opposite side to the leading foreleg.

A horse is said to be cantering 'false' and 'counter-lead' when he is cantering to the left with the off fore leading, or to the right with the near fore leading.

THE GALLOP

There are four beats to a gallop stride, so the gallop is termed 'four-time'.

The sequence of foot-falls is: with the right leg leading, left hind, right hind, left fore, right fore; with the left leg leading, right hind, left hind, right fore, left fore.

In gallop, the horse's outline should lengthen considerably. His stride should lengthen. He should cover the maximum amount of ground with each stride. He should continue to accept the bit and to remain in balance.

Aids to the Gallop

From the canter you should apply the aid for increased impulsion until the desired speed is reached (see 'The Aids', page 17).

The horse must continue to accept the contact of your hand, and you must take up the galloping position, maintaining a good contact with the hand and leg.

TRANSITIONS (Changes of Pace or Speed)

Transitions should be made with obedience and accuracy. They must be smooth and must flow one into the other without loss of rhythm, tempo or balance.

The horse should not alter his outline other than to adapt to the new pace or speed required. He should go clearly into the new pace without any shuffling steps.

The quality of the preceding pace influences the transition.

Aids to Increase Pace or Speed

If necessary, you should correct the quality of the pace as described under 'The Walk', 'The Trot' and 'The Canter', and then give the aid.

You must sit softly in the saddle, your body absorbing and feeling the movement of the horse and the alteration of the pace.

Your hands must accompany the movements of the horse's head and neck throughout. This is of particular importance in transitions, as any stiffening or fixing of your hands will result in resistance and loss of rhythm.

Aids to Decrease Pace or Speed

If necessary you should correct the quality of the pace, making sure that the horse is in balance and going with impulsion.

Throughout the transition you should sit softly and maintain the correct contact with the legs and reins.

Ask with a quick take-and-give of your outside hand for the alteration to the pace, i.e. canter to trot, trot to walk, while the legs maintain the necessary impulsion.

As the horse responds and the change of pace is accomplished, you may have to make some adjustments to the speed or the impulsion in order to correct the balance of the horse.

It is usual for an untrained horse to go too fast in the first few strides after the decrease of pace. The solution lies in improving the quality of the preceding pace.

Throughout the transition down, the horse must continue to accept the bit and maintain his outline and balance.

THE HALT

At the halt the horse must stand quite still and straight, his weight distributed equally over all four legs, fore and hind being in pairs abreast with each other. This is termed 'standing square'. Throughout the halt he must remain on the bit and balanced. He must continue to accept the bit. (He can be allowed to champ the bit quietly.) The rider must remain attentive.

The halt must not be abrupt. The steps preceding the halt should retain their correct tempo and speed.

Aids to Halt

These are the same as for aids to decrease pace.

During the halt you should maintain the contact of hands and legs, with both you and the horse remaining attentive to each other.

Avoiding stepping back Stepping back usually occurs when the horse is allowed to drift into halt without maintaining his impulsion.

If the rider maintains the impulsion while he rides forward into the halt and then pays particular attention to the contact of his legs and hands, the horse should stand at attention and be ready to go forward when asked.

The hand aids should be sympathetic and not forceful.

THE REIN BACK

The rein back is usually carried out from the halt. It has two steps to the stride and is almost in two-time.

The sequence of steps is: left hind and right fore together, right hind and left fore together.

The steps should be straight, active and unhurried, but of good length. The feet must be picked up and put down cleanly, with the horse maintaining his correct outline and remaining on the bit. He must not raise his head or hollow his back.

The rein back should not be attempted until the horse is going correctly in his basic paces and is able to balance himself through transitions and in halt.

Aids to Rein Back

From halt, you should maintain the contact of your legs and hands, the horse standing to attention. Ask for forward movement, but instead of allowing the horse to move forwards your hands should become slightly 'non-allowing'. Your body should remain upright and your back and hips supple. Force must never be used.

As the desired number of steps is accomplished, the horse

must be asked to go forward immediately. Indicate this with the appropriate leg aid and allow the forward movement with your hands.

CIRCLES AND TURNS

When moving on a curve, circle, or round a corner, the horse must bend his body uniformly from poll to tail to conform with the curve on which he is travelling *(figure 6)* i.e. the smaller the circle, the more acute the curve.

When on a curve, the footprint of the outside hind foot must be on the same line as the outside front foot.

The horse must be in balance and must not lose tempo or rhythm. His outline should not vary.

The horse's head and neck must not be bent outwards, nor should they be bent inwards more than the rest of his body.

The Aids to Circle and Turn

At all times you should keep your seat in the lowest part of the saddle, your hips parallel with the horse's hips, your shoulders parallel with the horse's shoulders, and your head turned to look in the direction in which you are about to go.

Your inside hand indicates the direction with a quick take-and-give.

Your outside hand controls the pace and allows the alteration in the bend of the horse's body. It also stops the horse's neck from bending too much to the inside.

Your inside leg maintains the normal contact with the horse's side on the girth, while the outside leg maintains normal contact behind the girth. If more impulsion is required, ask for it by giving a quick nudge with your inside leg.

3

BEYOND BASIC TRAINING

The Advanced Paces
Lateral Work

Advanced riding is only an extension of correct basic riding. The training of the horse and rider must be progressive. All horses are individuals and take a different amount of time to reach the various stages of training. It is wise to remember the saying, 'make haste slowly'.

Only when a horse is going in the correct way in all his basic paces can he be expected to proceed to more advanced work. This may be the time to introduce him to a double bridle.

10 *The trot, showing some collection (see page 36).*

To train the horse to a more advanced standard, firstly, the rider must be correct in his riding position and his application of the aids, and secondly, he must be a thinking rider.

With more advanced riding especially, the rider's body must be in balance and harmony with the movements of the horse. The aids must be clear, precise and accurate. You must be aware of every aid you give and train yourself to think before you act.

You must develop the ability to know instantly when your horse is going correctly in outline and pace. This is called 'feel'. It is easier for a rider to improve his riding on a trained horse which responds to correctly-given aids; it is easier for a horse to make progress when the rider is experienced. However, with good instruction it is possible for both horse and rider to make progress together.

11 *The trot, showing some extension (see page 37).*

You must never lose your temper; you must try to understand how the horse thinks and be aware of his mental weaknesses and physical difficulties.

You must be quick to praise and you must know when to correct.

Above all, you must learn to recognise the moment at which to stop each training session.

If the horse has not responded satisfactorily the rider should know whether it is because:

- The rider is at fault.
- The horse has not understood.
- The horse is unable to respond (physical weakness).
- The horse won't respond (disobedience).

More advanced riding requires a disciplined and thoughtful approach from the rider.

The Advanced Paces

A precise definition of the correct paces as laid down by the International Equestrian Federation is published by The British Horse Society's Dressage Group in their Dressage Rule Book. As small changes occur every year it would be wise for those interested in further training to read a current copy with care. The advanced paces, in simple terms, are set out below.

THE MEDIUM WALK

The medium walk, previously described in Basic Paces (page 25), is the pace between collected and extended walks.

THE COLLECTED WALK

The collected walk is the same as the medium walk except that:

- The speed is slower because the steps are shorter and higher. The joints of the hind legs and the forelegs are more active.
- The hind feet do not overtrack the prints left by the fore-feet.
- The head and neck are raised, and the head approaches the vertical position.

The rhythm and tempo do not alter.

This walk is difficult to ride correctly and should not be attempted too soon or for too long.

The Aids

The rider asks for collected walk by slowing down the speed and by controlling and maintaining the impulsion. Rhythm and tempo are maintained by the sensitive influence of legs and hands.

Common Faults

The rhythm may become irregular (broken). The horse may attempt to slow the tempo instead of shortening his steps.

THE EXTENDED WALK

The extended walk is the same as the medium walk except that:

- The speed of the walk is faster because the horse takes longer steps, covering as much ground as possible with each stride.
- The hind feet should clearly overtrack the print left by the fore-foot on the same side.
- The outline of the horse should appear to lengthen, with the head and neck stretching forward.
- More impulsion is required for extended walk than for medium walk.

The horse must remain on the bit.
The tempo and rhythm do not alter.
Before asking for extension you should first check the quality of the previous pace.

The Aids

The rider asks for extended walk by creating more impulsion, his seat remaining in harmony with the movements of the horse.

The hands maintain the contact and move with the stretching of the horse's head and neck.

Common Faults

The horse may lose balance and suppleness, and the steps

may become irregular. He may attempt to quicken the tempo of the walk instead of lengthening his strides.

THE WORKING TROT

The working trot, previously described in Basic Paces (page 26), is the pace between collected and medium trots.

THE COLLECTED TROT

The collected trot *(figure 10)* is the same as the working trot except that:

- The speed of the trot is slower because the steps are shorter and higher.
- The legs are more raised and their joints more active.
- The increased action of the joints of the hind legs makes the hindquarters appear lower, while the weight on the forehand appears to be lightened.
- The head and neck are raised, and the head approaches the vertical position. He does not track up.

The horse remains supple and flexible, with a swinging back.
The horse must remain on the bit.
The tempo and rhythm do not alter.

The Aids

The rider asks for the collected trot by slowing the speed and maintaining the impulsion.

Common Faults

- The horse may become tense and stiff in his back, thereby losing suppleness and activity.
- His steps may become irregular.
- The horse may attempt to slow the tempo of the trot instead of heightening and shortening his steps.

THE MEDIUM TROT

The medium trot is the pace between working and extended trots. It is the same as the working trot except that:

- The speed of the trot is quicker because the horse takes longer steps, covering more ground.
- There is more impulsion, the horse going forward with rounded and moderately extended steps.

- The outline of the horse lengthens with a lowering of the head and neck, and with greater engagement of the hocks.

The horse must remain in balance.
The rhythm and tempo must remain the same.

The Aids
The rider asks for the medium trot by creating more impulsion and then allows the horse to lengthen his outline and increase his speed by following the movement of the head and neck with his hands.

As the desired speed is reached, the rider controls the pace.

The hands must maintain the contact and follow the stretching of the horse's head and neck.

Common Faults
- The horse's steps may become irregular.
- He may attempt to quicken the tempo of the trot instead of lengthening his strides; this is termed 'running'.
- He may lose his balance and fall on to his forehand.

THE EXTENDED TROT

The extended trot *(figure 11)* is the same as the medium trot except that:

- The speed is faster because the horse takes longer steps, each stride covering as much ground as possible.
- The extended trot requires more impulsion than the medium trot.
- The hindquarters must be particularly well engaged and active to enable the horse to maintain the correct balance.
- The feet should come to the ground in a continuous forward movement and not stretch forward and then draw back.
- The outline of the horse should lengthen with a corresponding lowering of the head and neck.

The horse must remain on the bit.

The Aids
The rider asks for the extended trot by creating impulsion and, while maintaining the contact, allowing the horse to lengthen his outline.

The balance of the horse is maintained by the rider controlling the pace.

With all lengthening of strides the rider must create more impulsion in preparation for the increase of speed. The horse will lose balance if the rider attempts to lengthen the strides when insufficient impulsion is available.

Common Faults

These are the same as for medium trot.

Lengthened Strides

Sometimes a rider is asked to show some lengthened strides.

To do this, the speed of the working trot is increased by lengthening the strides to some extent. It is only necessary that some difference be observed.

It is more important for the horse to remain in balance, maintaining the same rhythm and tempo, than to show maximum lengthening.

THE WORKING CANTER

The working canter, previously described in Basic Paces (page 27), is the pace between collected and medium canters.

THE COLLECTED CANTER

The collected canter is the same as the working canter except that:

- The speed of the canter is slower because the steps are shorter.
- The hindquarters are more engaged and active, lightening the forehand and giving the shoulders greater freedom.
- The horse shows increased suppleness.
- The head and neck are raised, and the head approaches the vertical position.

The horse must remain on the bit.

The tempo and rhythm should remain the same.

The Aids

From the working canter the rider asks for the collected canter by slowing the speed and maintaining the impulsion.

Collected canter may be asked for from any pace.

Common Faults

- The horse may become tense and stiff in his back, thereby losing his suppleness and activity.
- He may lose the rhythm and tempo.
- He may lose the correct sequence of foot-falls.
- He may swing his hindquarters, thereby becoming crooked.

THE MEDIUM CANTER

The medium canter is the pace between working canter and extended canter. It is the same as working canter except that:

- The speed of the canter is quicker because the horse takes rounder and longer steps, the strides covering more ground.
- There is more impulsion.
- The outline of the horse lengthens and there is a lowering of the head and neck, with greater engagement of the hindquarters.

The horse must remain on the bit.
He must remain in balance.
He must maintain the tempo and rhythm.

The Aids

From working canter the rider asks for medium canter by creating more impulsion and then allows the horse to lengthen his outline and increase the speed. The contact must be maintained but not fixed. The rider controls the pace as the desired speed is reached.

Common Faults

- The horse may lose his balance and fall on to his forehand.
- He may quicken the tempo due to lack of impulsion.

THE EXTENDED CANTER

Extended canter is the same as medium canter except that:

- The speed is faster because the horse takes strides of maximum length, covering as much ground as possible with an increase of power.
- There is considerable impulsion emanating from the hindquarters and carrying the horse forward over the ground. The outline of the horse should lengthen with a corresponding lengthening of the head and neck.

The horse should remain on the bit.
He should maintain his balance.
He should maintain the rhythm and tempo of the pace.

The Aids

The rider asks for the extended canter by creating more impulsion and, while maintaining the contact, allows the horse to lengthen his outline and stride.

As with the extended trot, the horse will lose balance if the rider attempts to lengthen the strides before sufficient impulsion is created.

Common Faults

These are the same as for medium canter.

TRANSITIONS

In more advanced riding, transitions may be directly from any one pace to any other, in which case there would be no transition through an intermediate pace, i.e. canter to walk, or halt to trot.

Lateral Work

Lateral = sideways. The horse ceases to move with his hind legs following in the tracks of his forelegs, but proceeds down a straight line or on a diagonal, with fore and hind feet making separate tracks.

Most lateral work is in trot and canter, but some movements are in walk.

Some training exercises, such as turn on the forehand and leg-yielding, may be done before the horse has achieved collected paces, but otherwise some degree of collection is necessary before attempting lateral work such as shoulder-in, half-pass or demi-pirouette *(figure 13)*.

Throughout all lateral work the horse must remain on the bit in a correct outline, well balanced and going forward without resistance, maintaining the rhythm and tempo of the pace. Above all he must be relaxed; a tense horse cannot develop the suppleness that lateral work encourages and requires.

12 *Half-pass to the right (see page 46).*

MOVING OVER

This is the first exercise which a horse learns that involves lateral movement. He is taught how to move away from the leg. As a preliminary, it is useful to train the horse to move over in the stable by holding his head and lightly pushing him just behind the girth, while saying 'move over'. Progress to the same exercise when he is tacked up and outside. It is then logical for him to move away from the leg when the rider exerts some pressure either on or behind the girth, depending on the movement required.

TURN ON THE FOREHAND

This may be taught as soon as the horse understands the preceding exercise. While turns on the forehand, carried out equally on both reins, are a good way of teaching the first lessons in moving away from the leg, they should not be done too often as they do not encourage good forward movement, and are inclined to lighten the hindquarters rather than encourage their engagement.

This exercise is a test of a horse's obedience to the rider's legs and is useful training for opening and shutting gates. It is also a very valuable exercise from the rider's point of view, as it develops independent (unequal) use of his legs.

The rider prepares the horse for a turn on the forehand by walking and, if in a manège, by coming off the track by at least 1.5m (5ft).

The horse is halted squarely and immediately turned, head towards the track, up to 180°, with his hind legs crossing and thereby making a half circle round the inside foreleg on the side to which he is turning. The horse must then go forward at once into walk, trot or canter.

The sequence of walk steps must be maintained throughout. The horse should bend slightly in the direction in which he is turning. There should be no step backwards. There should be no resistance in the mouth or dropping of the bit.

The Aids

The rider asks for a turn on the forehand by halting squarely and immediately indicating the direction of the turn with the inside rein, while maintaining the contact of the outside rein to stop the horse gaining ground. The inside leg, just behind the girth, presses the quarters over, step by step. The outside leg behind the girth receives and regulates each step.

Common Faults

- Resistance due to not remaining on the bit.
- The horse may show too much bend in his neck in the direction in which he is going. This is corrected by maintaining the outside rein contact.
- The horse may step backwards, but he should not do so if the rider maintains the impulsion and keeps a good contact.
- He may walk a small circle. This is due to lack of control with the outside rein.
- He may lose the correct sequence of foot-falls. This is usually due to lack of impulsion.

LEG-YIELDING

Leg-yielding is carried out at trot or occasionally in walk, and is used to teach young horses early lateral work and an understanding of lateral aids. Collection is not required.

In leg-yielding, the horse moves on two tracks, i.e. forwards and sideways *(figure 13a)*.

His body is straight except for a very slight bend at the poll away from the direction in which he is going.

In the manège the horse remains parallel with the long side, but moves on a diagonal line.

This is a controversial movement, as some trainers believe that there are disadvantages in its use.

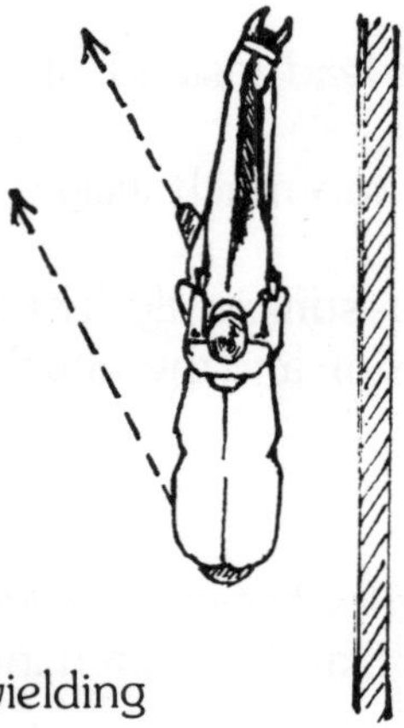

(a) Leg-yielding

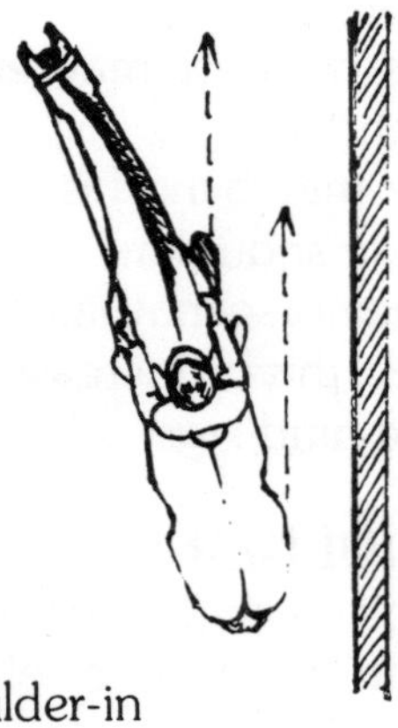

(b) Shoulder-in

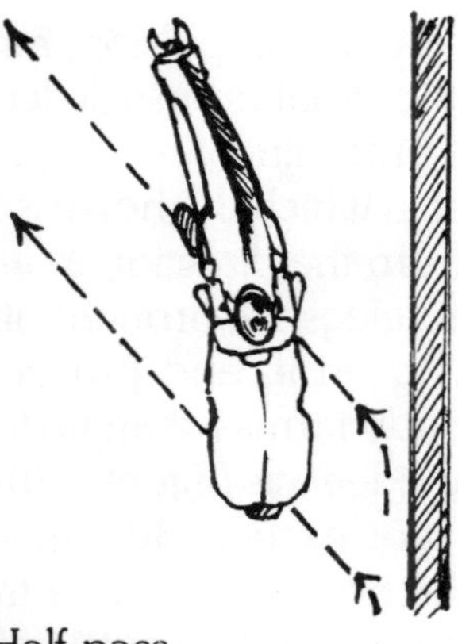

(c) Half-pass

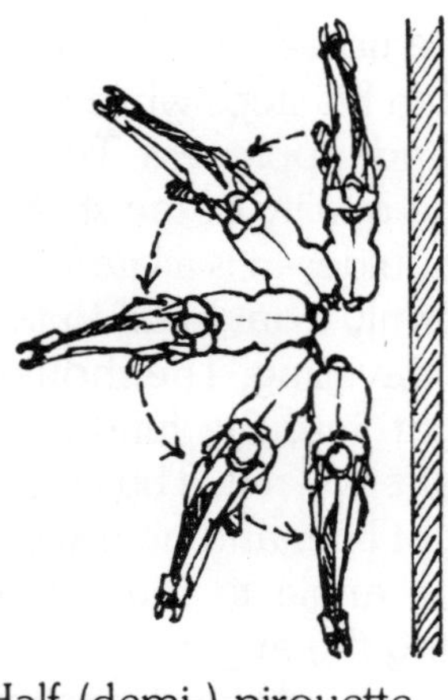

(d) Half-(demi-) pirouette

13 *Lateral work.*

The Aids

The rider asks for leg-yielding either down the long side of the manège with the horse's tail to the boards, or on a diagonal line.

When moving to the right, the inside leg (left), either on, or slightly behind the girth, moves the horse over; the outside leg (right) maintains the impulsion and the straightness of the horse; the inside hand (left), indicates the bend, and the outside hand (right) leads the way, controls the straightness and the speed of the forward movement.

Common Faults

- The horse may lose impulsion and rhythm, and fail to go forwards.
- His quarters may either trail or lead instead of remaining straight.
- He may bend his body and put too much weight on to his outside shoulder.
- The horse may fail to cross his legs sufficiently. The rider may correct this by increasing the impulsion and thereby the activity of the hind legs.

SHOULDER-IN

This is a valuable training exercise for horses that are already able to show some collection in trot, and it is often included in dressage tests.

The horse is ready to begin the exercise when he is able to shorten his steps while maintaining rhythm, impulsion, relaxation and suppleness. The quality of the shoulder-in will depend on the quality of the shortened trot preceding it.

Shoulder-in is a lateral movement in which the horse is bent uniformly from head to tail, away from the direction in which he is travelling. The shoulders and forelegs are brought off the straight line, the hind legs remaining on it, and proceeding straight down it. Thus the horse's body forms an angle to the straight line, and he travels partially sideways *(figure 13b)*.

The angle to the straight must not exceed 30°. In early training the angle is so slight that the horse is on three tracks with the inside hind leg stepping in the track of the outside front leg. The angle may be increased until the horse is travelling on four tracks with each leg making its own track. Shoulder-in may be done down the long side, down the centre line, or off a circle. The shoulders are always brought to the inside, off the track, i.e. right rein, right shoulder-in.

On completion of the shoulder-in, the rider may either ride forward on a curve across the school or straighten the horse by bringing the shoulders back on to the straight line. The former is less demanding and also more encouraging and rewarding to a young horse. The latter is used in competitions and is a greater test of obedience.

Shoulder-in is of great value, when carried out correctly, for increasing suppleness.

The Aids

The rider asks for shoulder-in by shortening the trot steps (see Collected Trot, page 36) possibly making a 10-metre (32-ft) circle before the corner at the beginning of the long side to encourage the correct bend. He then proceeds down the long side, bringing the forehand off the track as if he were beginning a second circle. At the moment when the shoulders and fore-legs leave the track, the rider's outside hand prevents the horse from gaining ground, and controls the degree of bend to the inside.

The inside leg maintains the impulsion and encourages the sideways movement, while the outside leg is a little behind the girth and prevents the quarters moving out.

The rider sits centrally in the saddle and looks between the horse's ears.

Common Faults

- The horse may lose impulsion and not remain on the bit.
- The angle of the shoulder-in may vary. This should be corrected by the rider making his aids consistent and finding a balance between hand and leg. This problem sometimes starts when a rider asks for too great an angle at the beginning of the movement.
- The horse may bend his neck too much, putting too much weight on his outside shoulder. This is usually caused by insufficient outside rein contact.
- The horse's quarters sometimes go out instead of the

shoulders coming in. You must pay particular attention to bringing the shoulders in at the beginning of the movement and controlling the swing of the quarters with your outside leg.

- Varying rhythm and differing angles on left and right reins are problems you must be quick to correct.

This exercise quickly loses its value if relaxation is lost.

HALF-PASS

In half-pass the horse bends uniformly throughout his body in the direction in which he is going. He moves on two tracks with his shoulders slightly in advance of his hindquarters. The outside legs cross in front of the inside legs *(figures 12 and 13c)*.

Half-pass is usually carried out either from the centre line to the long side or from the long side to the centre line.

It is important that the horse should continue to move forward actively, as well as sideways. He must maintain good rhythm and impulsion.

Half-pass is carried out from collected paces and should not be attempted until these have been achieved.

It may be performed either in trot or in canter, but it is easier for the horse in trot.

The Aids

The rider asks for half-pass from a collected trot or canter, either by turning down the centre or by moving off the long side. You always begin the half-pass in the same direction as the rein you are on, i.e. right turn down the centre, half-pass right.

Bring the horse's shoulders off the track as for shoulder-in, indicating bend and direction with your inside hand. Your outside hand prevents the horse making too much ground forward.

The outside leg behind the girth moves the hindquarters over, and the horse sideways. The inside leg maintains the impulsion.

The rider must look in the direction in which he is travelling.

Common Faults

- The horse may show the wrong bend and put too much weight on to the shoulder that is leading the movement.

- He may have his hindquarters leading or trailing. You must be aware of the correct angle and try to maintain it by judicious use of the correct aids.
- Insufficient forward movement may cause loss of rhythm and varying impulsion.
- The horse may lack collection.
- Too much bend can produce loss of freedom of the inside shoulder (cramped).
- Insufficient crossing of the legs indicates lack of suppleness and lack of engagement of the hind legs.

PIROUETTE

In a pirouette the horse makes a turn of 360°, with the forehand making a circle round the inside hind leg. The sequence of steps remains regular and active. It may be performed at walk or canter. At canter it is a most advanced and difficult movement.

A half-pirouette (demi-pirouette) is a turn through 180°.

During the entire movement the horse should remain on the bit and there should be no backward steps. The outline should be maintained with no resistance in the mouth. Throughout there must be a feeling of forward movement *(figure 13d)*.

The Aids

The rider asks for a half-pirouette (demi-pirouette) from walk by having a collected walk, or at least shortening the steps of medium walk.

The outside hand slows the speed, the inside leg maintains the impulsion and makes sure that the rhythm and tempo remain the same. The outside hand then stops the forward movement and at the same moment the inside hand asks for the turn. The outside leg, behind the girth, holds the quarters in position. On completing the turn the forward steps are resumed. The steps should appear unhurried, regular and clearly defined.

The horse should bend in the direction in which he is going.

Common Faults

- The horse may lose the sequence of foot-falls as a result of lack of impulsion or from tension.

- The hind foot may stick and not continue to 'march'. This is usually due to lack of impulsion.
- Stepping back may be the result of too strong an outside rein.
- 'Turning on the centre', i.e. not round the inside hind leg. This is almost certainly due to insufficient outside leg aid.

Progression

The correct training of any horse is dependent on the gradual development of both the horse's understanding of what is required of him (learning) and his physical ability (muscular development).

The importance of correct basic training and steady progression cannot be over-emphasised. It takes many years to bring a horse to an advanced level and it is a most worthwhile and rewarding task.

4

JUMPING

Analysis of the Jump • The Rider Jumping
Some Common Faults
Training the Young Horse to Jump
Towards More Advanced Jumping
Schooling Fences, Ditches and Water

Analysis of the Jump

It is important to understand exactly how a horse jumps. For this reason the jump has been divided into five phases:

1 The Approach
2 The Take-Off
3 The Moment of Suspension
4 The Landing
5 The Get-Away (or Recovery)

For the horse, these phases are described below. The corresponding phases for the rider are described on page 55.

PHASES OF THE JUMP—THE HORSE

Phase 1: the Approach

With hocks underneath him and actively engaged, the horse must be going forward in balance with impulsion and rhythm.

Success over the jump depends *largely* on a correct approach.

Phase 2: the Take-Off

• Before the moment of take-off the horse lowers his head and lengthens his neck, measuring up the fence and preparing for the spring.

• At the moment of take-off he re-balances himself by shortening his neck and raising his head. He lifts his forehand, folds his forelegs, and then propels himself into the air from his hocks.

14 *The five phases of the jump showing the correct position of horse and rider at each phase.*

Phase 3: the Moment of Suspension
While in the air the horse should round his back, stretching his head and neck outwards and downwards to the fullest extent. All four legs are now tucked up close to his body.

A horse is said to jump 'flat' when he fails to lower his head and neck, hollows his back and jumps like a deer.

Phase 4: the Landing
The horse straightens his forelegs and prepares to meet the ground. Once more he momentarily raises his head to re-balance himself.

His back should remain supple so that his hind legs can land and follow through in preparation for the get-away.

Phase 5: the Get-Away (or Recovery)
In the get-away stride, the horse's hocks should come well underneath him, in order to maintain a rounded outline. In show jumping particularly it is vital to consider this phase.

It is here that the approach to the next fence begins. Balance, impulsion and rhythm must be re-established as soon as possible. In the cross-country a well ridden, fast get-away saves vital seconds.

The Rider Jumping

THE JUMPING SEAT

Basically, the classic balanced seat (pages 10 and 11) always remains the same, but for jumping the stirrups are shortened, and thus the angles at the knees and ankles become more acute, i.e. more closed up.

A horse will only jump at his best when the rider's weight is over the horse's centre of gravity. Shortening the stirrups will help the rider to keep his weight there.

The Base of Support

When you shorten your stirrups, you will find that the base of support, that is, the whole thigh from the point of the knee to the hip joint, becomes wider. The broader the base, the more assured the balance *(figure 15)*.

Adjusting the Stirrup Length *(figure 16)*

As a practical test, remove your feet from the stirrups and let your legs hang long. Now try folding forward from your hips (from the top of your legs not from your waist). You will find it very hard to balance. Your lower leg will slide back and you may have to lean on your hands to catch your weight.

Shorten your imaginary stirrups—as short as those of a

jockey. Fold forward. Now you will find that balance comes easily because you will have a broad base of support.

Although some shortening of the stirrups will enable the rider to keep with the movements of the horse over fences and at speed, if the stirrups are *too* short the rider's legs will lose contact with the horse's sides; aids will become difficult to apply, and this will result in loss of control.

For your correct jumping length, try without stirrups to find a length which will enable you to fold forward in comfortable balance without losing your lower leg position, but long enough to have full use of your legs and back. Adjust the stirrups accordingly and close up the angles (hips, knees, ankles). Make sure that you remain in position, and *do not slide to the back of the saddle.* Stirrup length is entirely a matter for the individual rider. When finding yours, seek for comfort and balance.

Feet must be firmly in the stirrups which should lie across the ball of the foot with the rider's weight evenly distributed—not pressing on one side or the other (see page 10).

Hips, knees and ankles must remain as supple and flexible as well-oiled hinges. Knees and ankles act as shock-absorbers

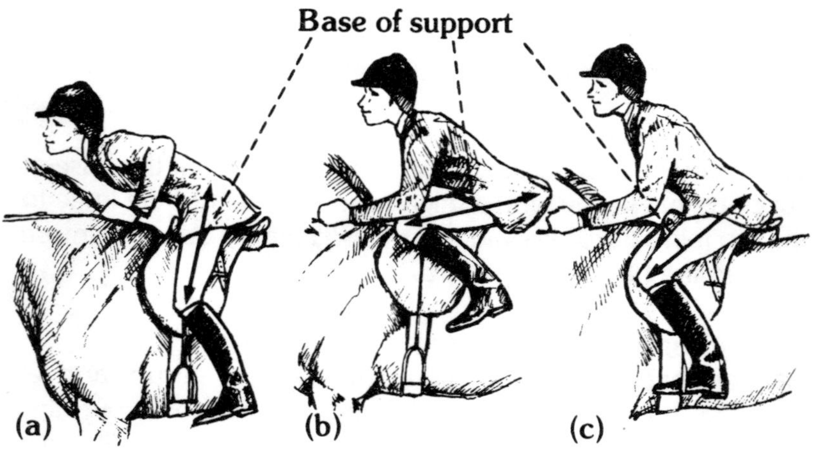

15 *The base of support.* **(a)** *Narrow base resulting in over-balancing.* **(b)** *Broad base – balance becomes easier.* **(c)** *The correct jumping length.*

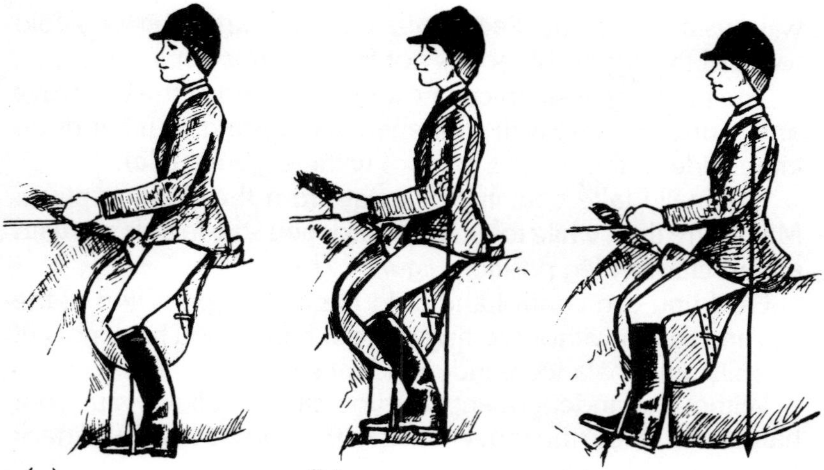

(a) (b) (c)

16(a) *The correct stirrup length on the flat.* **(b)** *The correct stirrup length for jumping.* **(c)** *Sliding back in the saddle after shortening the stirrups.*

and any stiffness makes it impossible to absorb the impact of the movement, especially on landing.

The lower legs rest softly in contact with the horse's sides, with the knees relaxed and well bent.

THE JUMPING POSITION

The jumping position is the position adopted by the rider's upper body (that is, from the hips upwards) as the horse takes off, and during the start of the moment of suspension (see Phases 2 and 3, page 55).

Practise this position at the halt. Fold forward from your hip joints (the top of your legs) with a flat back and with your spine in line with the horse's spine, not tipping to one side or the other. Look straight between the horse's ears, and keep your head up. Check that your hips remain supple, that your knees are bent, your heels well down, and that there is no over-gripping in your thighs (see Seat Faults, page 56).

Do not allow your lower leg to slip back.

In folding forward, feel as if you are putting your stomach towards the horse's withers with your shoulders following. This will *ensure* that you keep your back flat and that you fold forward from your hip joints not from your waist.

The jumping position can now be perfected at the walk, trot and canter. Practise with shortened reins resting your hands on either side of the horse's neck or withers *(figure 17a)*.

Gradually take your hands away from their resting place. Make sure that, while folding forward, you stay in balance with a correct lower leg position *(figure 17b)*.

Until you can control and hold the jumping position at the trot and canter without resting on your hands, you have not yet established a balanced, independent seat.

Without an independent seat you cannot 'allow' with your hands and follow the movements of the horse's head and neck (see page 11).

17(a) *The jumping position supported by hands resting on the horse's neck.*

17(b) *The jumping position with no support— hands away from horse's neck.*

Hands

When the reins are shortened, the position of the hands remains the same i.e. there should still be a straight line from the elbow through the hand down the reins to the horse's mouth *(figure 2)*.

PHASES OF THE JUMP—THE RIDER

Phase 1: the Approach

The rider sits securely but lightly in the saddle, so that his seat does not interfere with the balance, impulsion or rhythm of the horse.

The lower leg rests softly against the horse's side ready to give a quick aid.

Hands remain quiet, keeping a steady, light, contact. During the last three strides of the approach the contact must remain consistent; the horse's concentration must not be disturbed at this time.

The Beginner's Approach For a beginner to gain confidence and maintain balance, it is necessary to ride the approach with the upper body slightly forward, in anticipation of the horse's spring. For a novice rider to obtain this confidence it is essential to use a neck strap. The neck strap should be fairly tight. The rider tucks the first two fingers of one or both hands round the neck strap during the approach.

Holding on to the mane is a bad habit, as it encourages the rider to lean and to balance on the horse's neck, which nearly always results in a sudden loss of contact with his mouth (see Seat Faults, page 56). It is apt to produce a fixed hand over the jump.

Phases 2 and 3: the Take-off and Moment of Suspension

Seat The rider's body folds forward from the hip joints going with the movement of the horse. Your back should be flat, your head up, your eyes looking forward, and your weight evenly distributed (not leaning to one side).

Your legs should rest lightly against the horse's sides, with relaxed, bent knees, and heels supple and well down.

The rider's body should remain folding forward until the horse's hind legs have passed the highest part of the jump.

Hands While maintaining a light contact, hands and arms must follow the full head and neck movements of the horse, and should offer him as much rein as he requires, but no more (*figure 14*, pages 50–51).

Slipping the reins In the event of an error by horse or rider, the latter should be ready to open his fingers instantly to allow the horse to take the reins through his hands, re-establishing the contact as soon as possible on landing.

Phases 4 & 5: the Landing and Get-Away

As the horse prepares to land, the rider's upper body should swing rhythmically back into the upright position. The lower leg remains lightly in contact with the horse's sides, ready to pick up the stride for the get-away.

Care should be taken not to bump down on the horse's back when landing; it will interfere with his balance and recovery, and the bad memory of it may influence a future jump.

For procedure in learning to jump see *The Instructors' Handbook of The British Horse Society and The Pony Club.*

Some Common Faults

Many faults relate to each other, and sometimes one can lead to another. Stiffness and tension, often prompted by fear and apprehension, are the most common causes of faults, in both horse and rider.

RIDER FAULTS

Seat Faults

1 ***Complete loss of position*** This results from lack of ability or balance, or from leaning and collapsing on the horse's neck, and supporting himself on the hands.

2 ***Toes pointing downwards*** resulting in loss of position and control of lower leg. Aids thus become ineffective.

3 ***Behind or in front of the movement***

- Behind the movement—getting left behind as the horse jumps, cramping the back, often jabbing the horse's mouth.
- In front of the movement—leaning forward before the horse takes off, unbalancing and quite possibly inducing him to jump 'flat' because he will try to catch up with you (see The Jumping Seat, 'centre of gravity', page 51).

Both the above faults are the result of poor sense of timing and co-ordination.

4 ***Standing up in the stirrups*** with stiffened and straight knees instead of folding at the angles (hips, knees, ankles).

5 ***Dropping head and looking down*** Your head is the

heaviest part of your body; it is also your balancing factor. Dropping your head causes your back to round.

6 ***Rounding the back and leaning forward from the waist*** Unless your back is flat it is impossible to fold forward from the hip joints.

7 ***Over-gripping with thighs and knees*** This causes the rider to 'come up' in the saddle with lower legs well away from the horse's sides.

8 ***Tipping forward on to the knees*** This results in lower legs slipping up and back; loss of balance on landing; legs not in position to apply aids for the get-away.

9 ***Locked or stiff ankle*** Both this and (4) above, cause lack of flexibility in the joints and impair the effectiveness of the two main shock-absorbers—knees and ankles.

10 ***Leaning to one side or the other*** Tipping head sideways or looking back causes: a twist in the spine and uneven distribution of weight on the horse's back; loss of position and an inability to fold forward or to 'allow' with the hands. If very pronounced, this will cause a horse to jump crooked and/or drop a front leg.

11 ***Collapsing on landing*** The upper body over-balances, resulting in loss of control in the get-away.

Hand Faults

Many hand faults *(figure 18)* are the result of an insecure seat. Some of the most common are:

1 ***Catching the horse in the mouth*** Caused by getting behind the movement (Seat Fault 3), or by not slipping the reins if left behind unexpectedly.

2 ***Running hands up the mane*** and resting them high on the horse's neck.

Both the above faults can cause a refusal.

3 ***Hands fixed*** By not 'allowing', the horse is prevented from using his head and neck, often causing him to jump flat (see page 61) or, during the approach, to rush through frustration.

4 ***Dropping*** 'Throwing' the reins at the horse in the very last stride before take-off resulting in complete loss of contact. This unbalances the horse and causes him to lose confidence. It is often associated with Fault 2, above.

18 Some common hand faults

1 Catching the horse in the mouth

2 Running hands up the mane

3 Hands fixed

4 Dropping

5 Fiddling

6 Raising the elbows

7 Raising the hands as the horse takes off

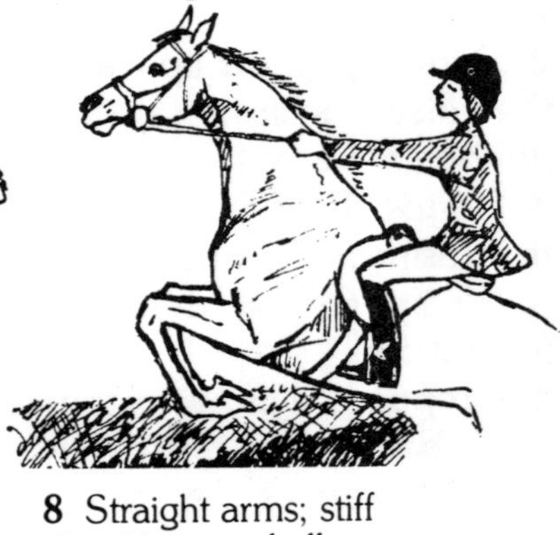

8 Straight arms; stiff shoulders and elbows

5 ***Fiddling*** This refers to busy, interfering hands that continually change the contact and muddle the horse during the approach (see page 55).

6 ***Raising the elbows*** Caused by a mistaken sense of 'allowing', or giving with your hands; in fact hands remain fixed, whilst elbows fly out sideways like wings.

7 ***Raising hands at the take-off*** Hanging on to the reins and trying to 'lift' the horse, which will restrict the freedom of his head and neck.

8 ***Straight arms, stiff shoulders and elbows*** Caused by tension in the rider. Your elbows act as cushions to your hands; to be effective they must be kept bent and flexible. Rigid shoulders can cause bad hands.

HORSE FAULTS

Refusing and Running Out

The main reasons are:

1 Bad riding
2 Pain
3 Fear
4 Disobedience
5 Fatigue

1 ***Bad riding***

- The rider, through lack of determination, fails to use the leg aids effectively and the horse immediately senses this indecision (as he will any fear or apprehension).
- Jabbing in the mouth, bad hands, not 'allowing', or loss of balance by the rider.
- Bad presentation and approach to the fence.
- Over-jumping which may cause the horse to become sickened.
- Dropping (see page 57). This is one of the most common rider faults for refusing in a trained horse.
- Over-checking—lack of impulsion and rhythm.
- Over-riding—unbalancing and driving the horse on to his forehand. Going too fast from too far out.
- Fiddling in the last strides of the approach. Constantly putting the horse wrong (see 'mistrust', page 68) and disturbing his concentration.

2 ***Pain***

- Lameness caused by pain in legs or feet.
- Discomfort in the mouth as a result of over-bitting or jabbing.
- Pain or tension in the back due to the rider bumping on the loins.

3 ***Fear*** caused by:

- Over-facing (asking a horse to jump beyond his ability or stage of training).
- Loss of nerve, or the recent memory of an unpleasant experience at a previous fence.
- Lack of confidence in the rider.
- Slippery going and bad take-off.
- Unsuitable obstacles.

4 ***Disobedience*** caused by:

- Lack of correct training.
- Ineffective riding.
- Nappiness (wilful disobedience to the rider's aids).

5 ***Fatigue*** caused by:

- Lack of consideration by the rider, i.e. going on for too long or not knowing when to stop.
- Lack of fitness or condition.

Rushing

The main causes are:

1 Bad riding.
2 Incorrect training.
3 Fear or pain.
4 Frustrating the horse by restricting, i.e. over-holding during the approach.

'Putting in a Quick One' (Putting in an unnecessary extra stride at take-off) and **'Propping'** (A sudden hesitation in front of the fence).

The main causes of both are:

- Incorrect riding in the approach.
- Not going forward, losing impulsion and rhythm.

- Losing balance or rhythm through going too fast too far out (see page 59).
- Dropping the contact at the last moment before take-off.
- Lack of scope or of muscle fitness in the horse.
- Soreness or pain, usually in the feet.
- Temporary disobedience or lack of confidence.

Jumping Flat

In this case the horse does not round or bend his back and therefore does not use himself to the best of his ability (see Phase 3, page 50).

The main causes are:

- Incorrect training or bad riding.
- A stiff or an injured back.

Some horses have a natural tendency to jump flat. They can be helped to overcome this by correct progressive training.

PREVENTION AND CURE

Basically, the cure for any fault is first to analyse it and then to diagnose the cause. Once the cause is established, common sense, experience and knowledge in training should cure the trouble. For the young rider it is almost impossible to accomplish this alone. Wrong analysis and faulty attempts at correction may do more harm than good. Help should be sought from an experienced instructor, if necessary in co-operation with a veterinary surgeon.

Because every horse and rider is entirely different, the solution to each problem is different; you cannot generalise and there is no copy-book answer. If there were, the assistance of an experienced instructor would never be required; you could simply look it up in a reference book.

Training the Young Horse to Jump

Horses are creatures of the plains. In their natural state, they do not normally jump. Therefore they must be taught. In the very early stages of training, while out riding, the young horse should be encouraged to pop over small natural obstacles which happen to be in his path (logs, fallen branches or small

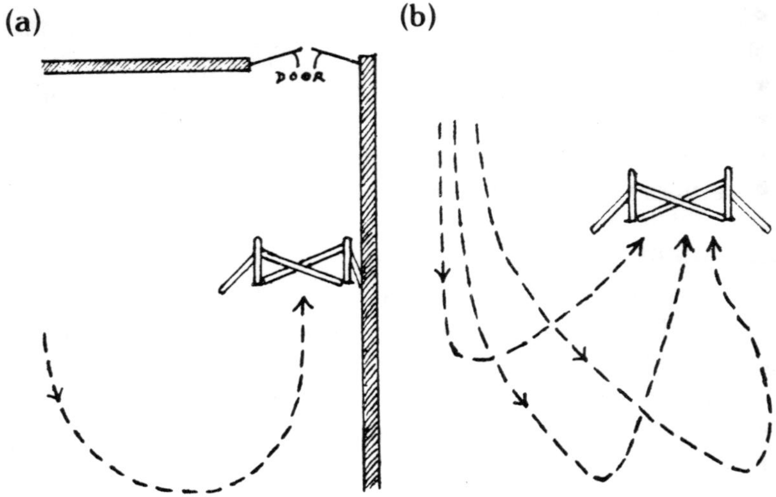

19(a) *The correct approach to a fence.* **(b)** *Three incorrect approaches to a fence.*

ditches). Approach in trot, preferably with an experienced companion to instil confidence and give a lead.

The rider, however good he may be, should always make use of his neck strap. Young horses are unpredictable; a jab in the mouth will long be remembered and feared.

It is vitally important to make the first jumps as attractive as possible in order to keep the horse relaxed and happy. A small obstacle with a wing can be introduced in a field, school or arena. Placed beside the wall or fence and towards the exit, this will help to keep the horse straight and will encourage him to go forward to his jump *(figure 19)*.

INTRODUCING THE FIRST TRAINING JUMPS

Horses are individuals and the length of time taken by each one to reach a particular stage of training varies.

Stage 1

- Begin by walking the horse over a single pole placed on the ground between two wings and against the side, not in the middle, of the school or field.
- Break into a trot, coming in on a wide semi-circle. In schooling, always approach on an even arc to ensure that the horse is in balance, with his hocks engaged *(figure 20a)*.

Stage 2

The poles are raised to small single cross-poles *(figure 20b)* (see Construction of a Basic Schooling Fence, page 83).

When these can be jumped with ease on either rein, small double cross-poles should be introduced—about 30cm (1ft) high and 30–60cm (1–2ft) wide. This fence will encourage the horse to jump in the correct style, bending and rounding his back from the start. Cross-poles assist a young horse to jump straight, guiding him towards the middle of the fence.

Practise jumping on either rein, gradually raising and widening the fence as the horse gains confidence. Never over-face him; *keep your fences small.*

For Stages 1 and 2 and for all young and inexperienced horses, *approach at the trot* because:

- The slower the pace the greater the control.
- The horse will arrive right for the take-off zone (a trot stride being less than half the length of a canter stride).
- Unhurried and in balance, the horse will have time to jump, and can 'use' himself and have no reason to be frightened. Fear is one of the principal causes of rushing or refusing (see page 60).

Do not proceed to Stage 3 until your horse is calm and relaxed. Tension causes bad style in jumping (see Relaxation, page 75).

Stage 3

Change the cross-poles to a simple staircase *(figure 20c)* (see Construction of a Basic Schooling Fence, page 83). Approach from the trot on a semi-circle.

Gradually, as the horse gains confidence, he can be allowed to break into a canter in the last few strides of the approach. Should he rush or jump flat, return to the approach in trot until the fault has been eliminated.

Stage 4

The horse should now be ready to jump a small double.

Young horses should be introduced to doubles as early as possible. Provided the fences are small and well-built, and the distances correct, they will encourage good jumping, help to train the horse's eye and teach him to think and to concentrate. Keep distances short—over-long distances tend to make

20 *Six stages in training the young horse to jump.*

(a) **(b)**

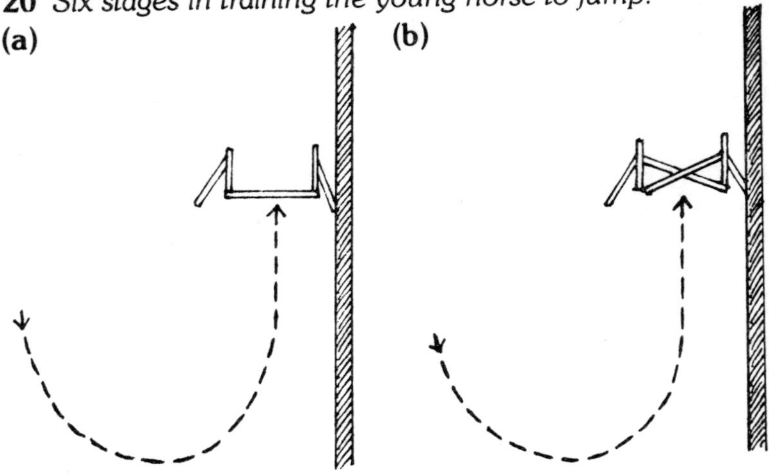

Stage 1 Approach in trot **Stage 2** Approach in trot

a horse jump flat (see page 61).

Start with a two non-jumping stride double (see page 86) with a staircase fence as the first element, and a simple upright (one single pole with a ground-line pole or dropper) roughly 9m (30ft) away as the second element *(figure 20d)*. Approach in trot for this distance.

Stage 5

- Widen the double to approximately 10–11m (33–36ft) and jump from the canter, only breaking into canter in the last few strides *(figure 20e)*.
- When the horse is completely confident over the two non-jumping stride double, close up the double for a one non-jumping stride. The distance will depend on whether the fence is jumped by a horse or a pony and on the *pace of the approach:* 5.5-6.4m (18-21ft) at the trot; 6.4-7.3m (21-24ft) at the canter.

Stage 6

Small fences of differing types can now be placed away from the side of the school or field, some on the diagonal; always with a true ground-line (see page 80) and of solid appearance *(figure 20f)*.

(c)

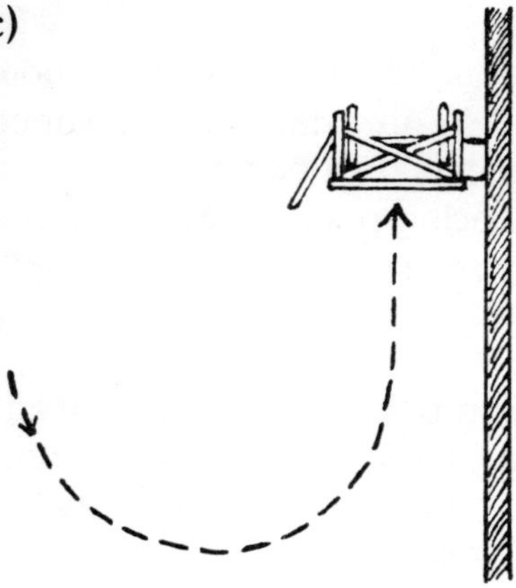

(d)

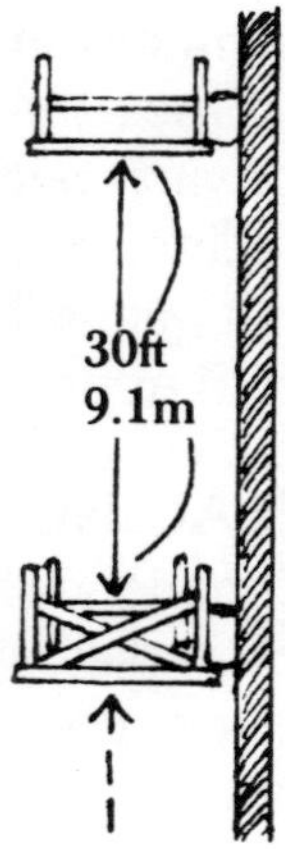

Stage 3 Approach in trot and, later, in canter

Stage 4 Approach in trot

(e)

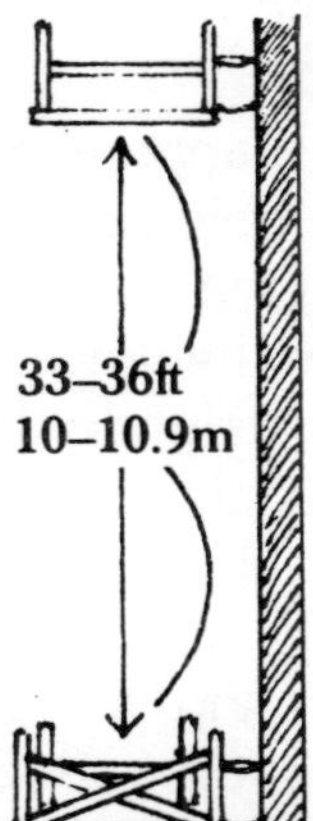

(f)

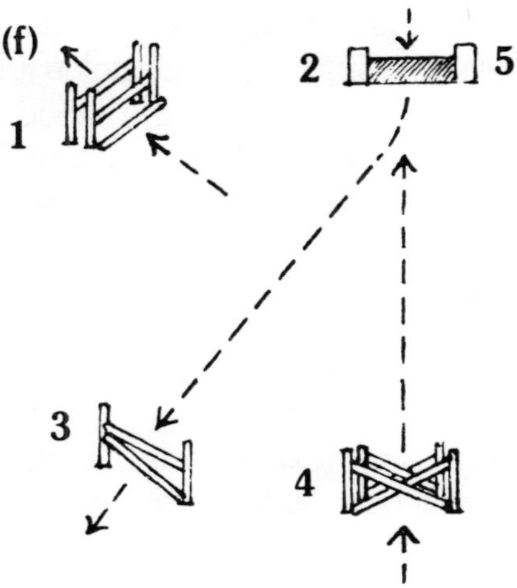

Stage 5 Approach in canter

Stage 6 Approach in canter

Ensure a good approach to each fence by making an even, rounded turn into the obstacle. Thoughtless riding of the track, resulting in square or pointed turns, unbalances the horse and makes him lose rhythm and impulsion (Centre Line, page 74).

If correctly trained from the start, a horse should *never know how to refuse.* However, should a refusal occur, the fence must be *lowered at once*, and be kept very small until confidence, calmness and obedience return.

Once the horse will jump small fences of all types, going forward between leg and hand in an established, balanced outline, then more advanced training can begin.

If take-off problems arise, a placing-pole can be introduced from approximately Stage 3 onwards.

PLACING-POLES

Experience has shown that placing-poles should not be used in the earliest stages of training. They tend to muddle and confuse a very young horse. However, once the horse has mastered small jumps placing-poles are invaluable.

A placing-pole is a pole not less than 10cm (4in) in diameter and about 3m (10ft) long, set on the ground approximately 2.7m (9ft) from the base of the fence on its take-off side. Its purpose is to assist a horse to arrive right for take-off, to encourage him to lower his head and neck, and to help him towards a smooth, comfortable jump.

At 2.7m (9ft) from the fence, the approach should be in trot *(figure 21a)*. Adjustment of the pole nearer to or further from the fence will assist in achieving the correct take-off.

If the horse stands off too far from the fence, then the pole should be brought in closer. If he gets under his fences and needs to be encouraged to stand back, the pole should be rolled further away: 15–45cm (6in–1½ft) in either direction should be sufficient.

For an inexperienced rider learning to jump with the approach in canter, a placing-pole 5–6m (18–20ft) from the fence will ensure a smooth and uniform jump *(figure 21b)*.

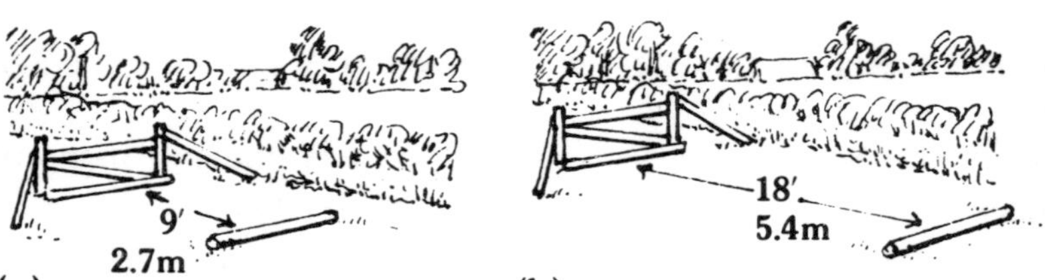

(a) **(b)**

21 *Placing-pole.* **(a)** *With the approach in trot.* **(b)** *With the approach in canter.*

Towards More Advanced Jumping

THE HORSE

The advanced horse must be trained to:

- Respond instantly to the aids and go forward from the leg immediately when asked.
- Shorten and lengthen his stride.
- Remain in balance through turns and corners without losing impulsion or rhythm.
- Change direction in balance without becoming disunited (a very common fault in novice jumpers).

All the above must be taught on the flat. The importance of ground work cannot be too strongly emphasised; for the competition horse it is essential. Success in jumping depends largely on the correct approach (see page 49).

THE RIDER

Before starting jumping practice, at least 20 minutes should be spent on ground work, in order to ensure that the horse is supple, obedient, in tune with his rider and well warmed up.

In more advanced jumping the rider can assist the horse to arrive right at the fence, i.e. reach the correct take-off zone (see page 82). He can regulate the impulsion and length of stride before an obstacle, and thus ensure that the horse approaches the fence with increasing momentum. Impulsion must *not* be confused with speed. It is a common fault to approach a fence far too fast from too far out, losing rhythm and balance, over-riding and giving the horse no time to jump (see Bad Riding, page 59).

Any adjustments must be made at least three strides before take-off. During the last three strides, the rider, while maintaining impulsion and rhythm, should *not* disturb the horse in any way. Here the horse must be allowed to concentrate on the jump and be given a chance to make his own adjustments if necessary.

It is the art of arriving right at the obstacle which makes for good clean jumping. This can only be achieved through constant practice, with a good natural eye, on a well-trained horse. Some riders have more talent than others, but with training and practice an eye can be improved. Unless the rider is certain that he can be accurate, it is better to keep the horse going forward in balance, with impulsion and rhythm, maintaining contact with legs, seat and hands, and leaving the horse to make his own adjustments.

A horse continually put wrong will soon learn to mistrust his rider and this will result in apprehension, rushing, 'putting in a quick one', or in stopping.

The advanced rider should approach the fence sitting in an upright position, with his seat-bones in light contact with the saddle. Only in this position will he be able to feel instantaneously any resistance which, unless immediately corrected, could lead to loss of impulsion, a refusal or run-out. This position gives the rider full power to carry out the delicate operation of shortening or lengthening the stride while retaining balance and rhythm with his horse.

TROTTING-POLES AND CAVALETTI

When used for canter exercises or grid work, cavaletti piled together to form a jump, have proved *very dangerous*. When hit they are liable to roll and become entangled in the horse's legs, bringing him down. They have caused serious accidents and injuries, and should never be used for jumping. True cavaletti, that is poles with a wooden cross bolted at either end, should only be used for trotting exercises, and should be set at their lowest. Always use heavy poles that cannot be easily dislodged.

Trotting-poles and cavaletti should never be more than 15cm (6in) high. If set higher than this, they cause tension and loss of rhythm and will encourage 'hopping'.

THE USE OF TROTTING-POLES

Far more satisfactory than cavaletti are poles placed on the ground—i.e. trotting-poles. They are more versatile, less clumsy, and safer.

Trotting-poles are heavy poles not less than 10cm (4in) in diameter and 2.4–3m (8–10ft) long *(figure 22)*. Three useful ways to prevent them from rolling and to give height when necessary are:

- To set each end in a log or wooden block approximately 0.6m (2ft) long.

22 *Types of trotting-pole.*

- To fix them in breeze blocks.
- In an arena, to heap sand on either side at each end.

Value to the Horse

As a *discipline* trotting-poles:

- Teach obedience
- Develop balance and rhythm.
- Teach mental and physical co-ordination, calmness and concentration. They help to develop the horse's natural eye.
- Develop the correct muscles and help to supple the horse by:

  Lowering the head and neck.
  Rounding the back.
  Engaging the hocks.
- Help to regulate and establish the stride.
- Teach the horse the correct approach to a fence, going forward with a rounded outline in balance, with rhythm and impulsion.

Value to the Rider

Trotting-poles teach and develop:

- Balance.
- Rhythm.
- Timing, 'feel' and co-ordination of aids.
- The rider's eye for distances and placing.
- A knowledge of how to ride a correct track (the path taken between the fences and in the approach while jumping a course).

Trotting-Poles in Rising Trot

Trotting-poles should be ridden in *rising trot*, firstly to encourage the horse to relax and 'swing' his back; secondly to develop the rider's sense of rhythm. (The sitting trot may cramp and flatten the horse's back and will not assist his sense of rhythm – see pages 24 and 72.)

Approximate Trotting-Pole Distances

Pony: 1.2–1.3m (4 – $4\frac{1}{2}$ft) apart.
Horse: 1.3–1.4m ($4\frac{1}{2}$ – $4\frac{3}{4}$ft) apart—sometimes 1.5m (5ft).
When a distance of 2.7m (9ft) between poles is used (see Stage 2, below) this is suitable for both horses and ponies.

Trotting-poles should be set just off the track so that they can be approached on either rein or by-passed.

The turn into them must not be sharp—minimum 15m (49ft) arc.

Never use fewer than three poles—two will encourage the horse to jump instead of to trot.

Procedure

Stage 1

Place trotting-poles far apart, at random around the arena (*figure 23a*).

Method: It is important to *walk* over the random poles until the horse becomes completely relaxed and at ease, eventually walking on a loose rein.

Pick up the contact quietly, trot on and trot over the poles, changing the rein at intervals. Make sure that the quality of the trot remains the same (see page 26) and that the horse does not try to rush or jump.

If problems arise, return to walk.

Stage 2

Lay trotting-poles out as shown below (*figure 23b*).

Method: Start by walking over the three poles set 2.7m (9ft) apart. Never walk over poles 1.3m ($4\frac{1}{2}$ft) apart—they are the wrong distance for the walk stride and will only upset and confuse a young horse. Use poles 1.3m (4½ft) apart for trotting exercises *only*.

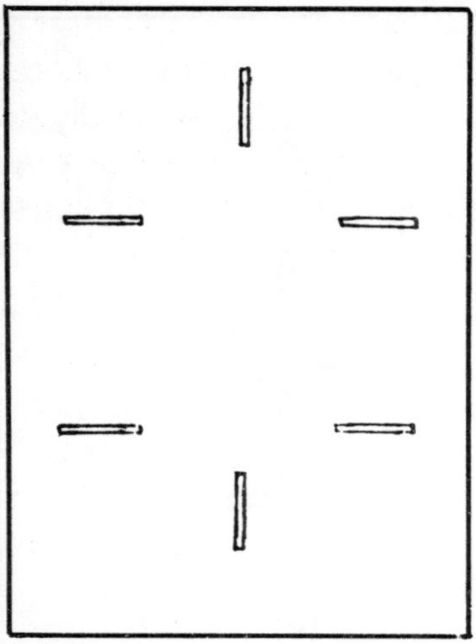

(a)

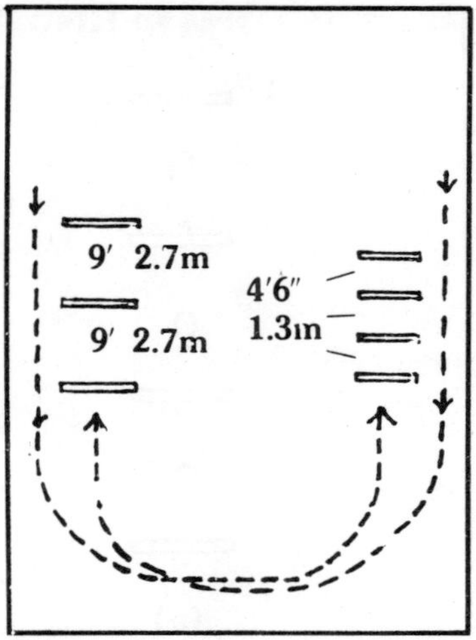

(b)

23 *Layouts for trotting-poles.* **(a)** *Set at random as for Stage 1.* **(b)** *Placed at distances of 2.7m (9ft) and 1.3m. (4½ft) apart as for Stages 2 and 3.*

Stage 3

When the horse is completely calm and relaxed at the walk over the 2.7m (9ft) poles, break into *rising* trot. Check the quality of the trot (see page 26) and proceed over the 2.7m (9ft) poles, coming in on an even, well-rounded arc to ensure balance and rhythm (see The Centre Line, page 74).

Change the rein and repeat.

Stage 4

Dispense with the 2.7m (9ft) poles.

Using a full circuit of the arena to re-establish the trot, proceed over the 1.3m (4½ft) poles. Repeat on either rein. Do not hurry; concentrate on *rhythm*.

Distances between poles are approximate, and will vary according to the stride and ability of the horse.

Checking Distances, Balance and Rhythm

When it is certain that the horse is working correctly, to check that the trotting-poles are set at the right distance the instructor should ensure that the hind foot falls exactly midway between the poles *(figure 24)*.

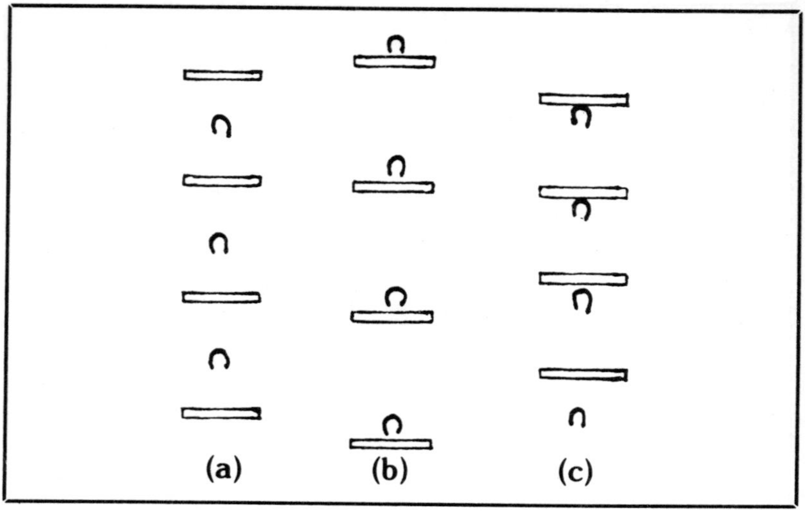

24 *Checking that trotting-poles are correctly placed.*
(a) *Correct foot-falls.* **(b)** *and* **(c)** *Incorrect foot-falls.*

In the final stages, to check that balance and rhythm are firmly established, add one more pole on to the 1.3m (4½ft) line, at a distance of 2.7m (9ft) away. Ride on either rein, starting with the 2.7m (9ft) space at the far end. Check that the horse maintains his rhythm even within the 2.7m (9ft) space where there is no pole on the ground to guide him.

While working over trotting-poles you *must* look up and *onwards.* Looking down, dropping your head or tipping forward will unbalance the horse, inviting him to jump or rush (see 'centre of gravity', page 51).

Rhythm Some riders find it hard to acquire a sense of rhythm. If so, it is helpful to repeat out loud a jingle such as, 'Monday, Tuesday, Wednesday', etc, during the approach and over the trotting-poles.

Rhythm is essential for all successful riding.

Alternative Exercises

As an alternative exercise, trotting-poles can be laid out on the diagonal.

The value of this exercise is to teach the rider to maintain impulsion and rhythm on the corners, while riding a correct track (see Value to the Rider, page 69), and to become aware

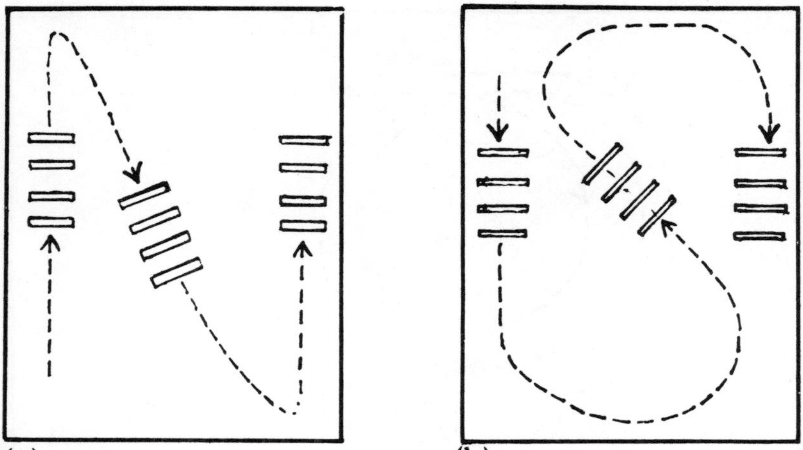

(a) (b)

25 *Alternative exercises with trotting-poles on the diagonal.* **(a)** *An incorrect track – sharp, angular turns.* **(b)** *A correct track – smooth, rounded turns.*

of and practise the use of the centre line. The track should flow like a river, with no pointed, sharp or square turns. Smooth, well rounded turns promote balance *(figure 25)*.

Warnings and Corrections

Before moving from one stage to the next, it is essential to perfect each exercise. If things go wrong, start again from the beginning.

Incorrect, hurried, trotting-pole work has *no value*.

If the horse persistently rushes or jumps the poles you should adopt one of the following alternatives:

- Walk the horse up to the line of 1.3m (4½ft) poles, halt, pat him, turn away left; walking in a half circle, walk up again; halt; establish the halt; turn away right; walk up again; halt; turn away. Continue to walk, halt, and turn away until the horse is calm and obedient. Then walk right up to the trotting-poles, start rising and gently ease to the trot over the first pole; continue down the line in trot. Return to walk and repeat.
- If the above correction has not been successful, trot in an inward circle towards the line of poles, by-passing the first

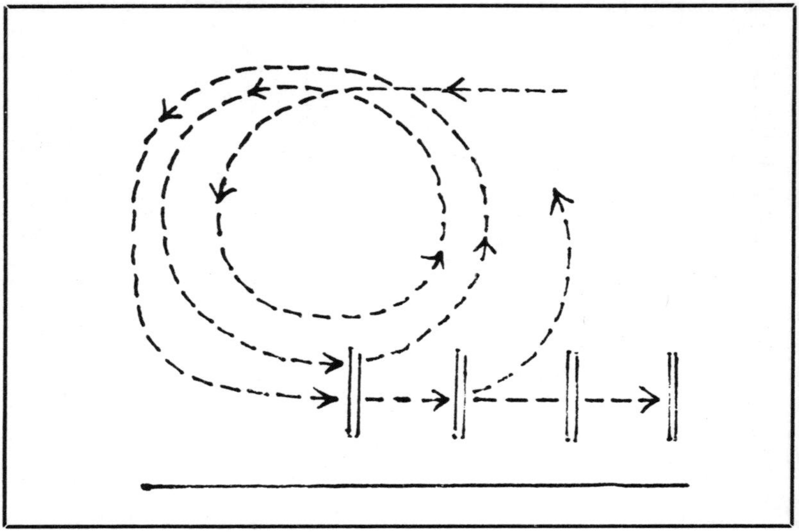

26 *An alternative exercise: trotting on an inward circle past the first pole (see Warnings and Corrections).*

pole. On the second circuit, trot over the first pole only. Continue circling, taking in one more pole each time as the horse becomes calm and obedient, and finally trotting down the entire line of poles *(figure 26)*.

The Centre Line

In riding a track, on the turns and during the approach to the fence, the rider must be aware in his mind's eye of the centre line. This is an imaginary line running along the ground straight through the centre of the fence or combination of fences. The rider should come in to meet the centre line on an even, rounded turn. Over-shooting the centre line and having to make an 'S' turn back on to it will cause your horse to lose balance, impulsion and rhythm in the approach *(figure 27)*.

Taking a short cut on to the centre line will not promote balance, and the horse will meet the fence at an angle. This will encourage him to run out because, especially in combination fences, he will see a way out and probably take it.

During the approach, when coming on to your centre line, especially in combinations, sight the fences so that they are straight in the horse's line of vision, one behind the other.

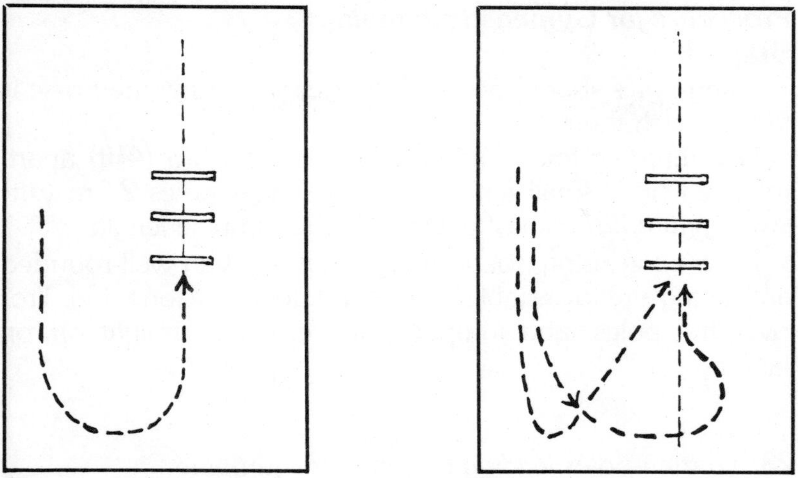

(a) (b)

27 *The centre line.* **(a)** *The correct approach on to the centre line.* **(b)** *Incorrect turns on to the centre line.*

GYMNASTIC JUMPING

Relaxation

A horse cannot jump really well unless he is relaxed. Gymnastic exercises if carried out correctly will promote relaxation and the horse will have no reason to feel fear or apprehension.

Trotting-Poles Leading to Gymnastic Jumping

Do not attempt to start gymnastic jumping with trotting-poles until your horse is proficient in all his trotting-pole work and has mastered small fences, including doubles, with a placing-pole (see page 66).

For more advanced work, trotting-poles and fences should be set down the centre of the arena, if space permits, to allow for the approach on either rein. The arc of the turn into the fences must not be less than 15m (49ft) in diameter. The horse should not be asked to bend more than he is capable of bending. A tight arc will unbalance him and make him lose rhythm unless he is very supple.

Fences preceded by trotting-poles down the side of the school or against a hedge will help a novice horse, but he will be able to come in on one rein only (see page 62).

Procedure for Gymnastic Jumping

Stage 1

Trotting-poles should be used on their own until the horse is familiar with them.

Use three or four trotting-poles placed 1.3m (4½ft) apart, followed by a small fence of double cross-poles 2.7m (9ft) away *(figure 28)*. Adjust your stirrups to jumping length.

Proceed in rising trot, coming in on an even well-rounded arc having already established the trot down the long side. Trot over the poles, and jump the fence. *Keep straight on* on landing.

28 *Trotting-poles leading to gymnastic jumping.*

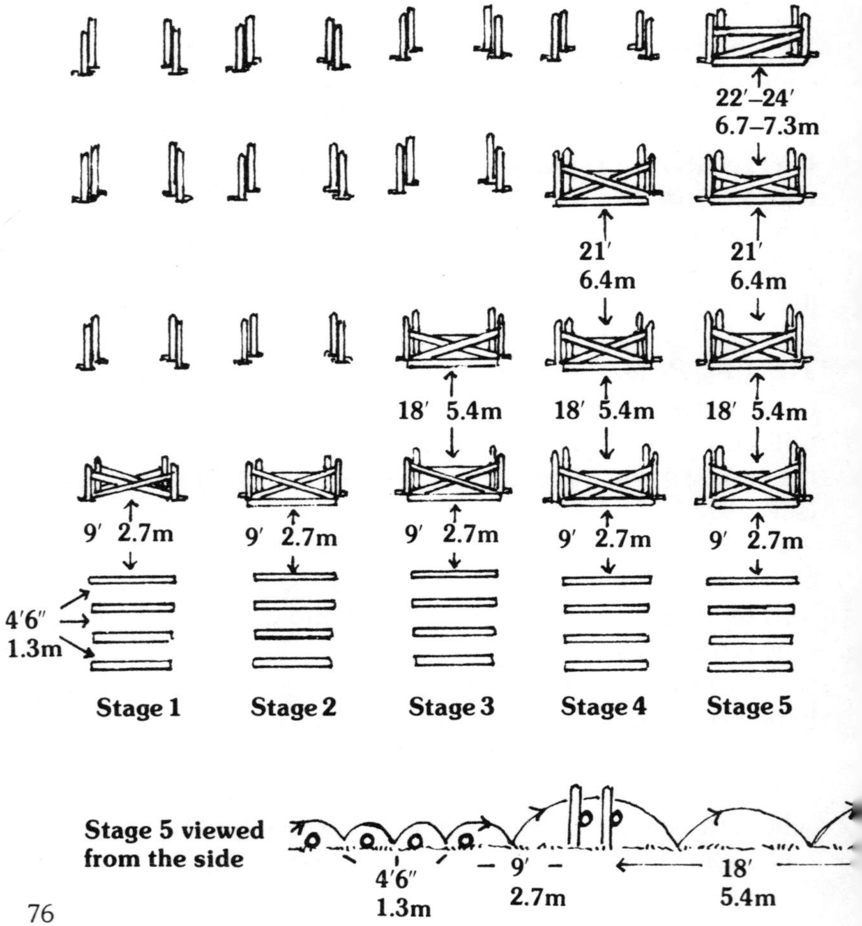

Stage 2

Trotting-poles remain the same as for Stage 1. Alter the fence to a basic schooling fence of simple staircase design (see page 83). In the early stages, keep fences small—not more than 0.6m (2ft) high and wide.

Proceed as in Stage 1.

Stage 3

Add a second fence of similar construction (simple staircase) as in Stage 2, 5.4m (18ft) from the first. Measure the distance from the back of the first fence to the face of the second fence. A distance of 5.4m (18ft) between fences will ride correctly for one non-jumping canter stride between the two elements *provided the approach is in trot (figure 28)*.

Proceed as in Stages 1 and 2.

The first element should remain a simple inviting staircase fence. As training progresses, the second element can be widened, raised and eventually become a parallel.

Stage 4

Add a third fence 6.4m (21ft) from the second fence. Start by using a simple ascending oxer. Vary construction as training proceeds. Ride as in Stages 1, 2 and 3, coming in on an even, rounded turn, trotting over the poles and jumping down the line with a one non-jumping canter stride between each element *(figure 28). Keep straight on* on landing.

Stage 5

Add a fourth fence—a simple upright—consisting of one pole with a dropper (see page 84) approximately 7.3m (24ft) from the third fence. The upright must *not* be changed to a parallel or a spread until confidence has been completely established over all four elements. A faulty attempt at an upright fence will not cause fear or pain.

The distance between fences 3 and 4 will vary according to ability and training *(figure 28)*.

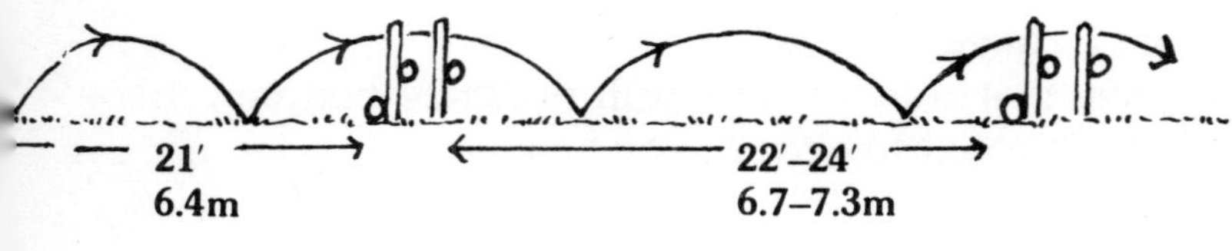

For ponies and short-striding horses with little scope a distance of 7.3m (24ft) between the last two elements will be too long. They will not be able to make it without an effort and may be tempted to 'put in a quick one'.

Shorten the distance to 6.7 or 7.0m (22 or 23ft) for them. Try 6.7m (22ft) first.

Effort promotes tension (see Relaxation, page 75). Too long a distance will flatten and spoil the jumping style.

HEIGHTS AND SPREADS

The four obstacles described in Stages 1 to 5 can remain the same size or become gradually and progressively larger. In combinations, never have one element lower than the previous one since this will encourage 'diving'. (This will not apply to later training, once the jump has been firmly established.)

The height and spread of fences depends on training and ability. Spreads in combinations can cause difficulties.

Never over-face. Use small fences for schooling until the horse is confident and happy.

Small ponies and very novice riders should not attempt gymnastic jumping with spreads. They should use only upright fences.

Alternative Gymnastic Exercises

Once the horse has mastered simple gymnastic jumping as in Stages 1–5, striding-distances and types of fences can now be altered as a variation for each exercise, using a bounce and one or two non-jumping canter strides between the elements (see Bounce Fences).

From now onwards trotting-poles should not be necessary as an introduction to the obstacles. Use only a placing-pole (see page 66).

Distances must be correct and will depend largely upon the pace of approach, the types of fences used, ability and progress in training.

Bounce Fences

A bounce, that is two fences with no stride between them, should be introduced early during training.

Approaching in trot, bounce fences should be from 2.7–3.3m (9–11ft) apart, and in canter from 3.3–4.3m (11–14ft)—a little less for ponies.

Bounce distances should be incorporated into gymnastic jumping, preferably as the first and second elements.

Unless the horse is thoroughly experienced and obedient, do not attempt a bounce at the end or middle of a gymnastic jumping line where the preceding distances have been for one or two non-jumping strides. (The horse will try to jump the bounce.)

Warnings

- Gymnastic exercises should not be attempted unless the rider has sufficient knowledge and experience. Without this, the result will be a disaster, especially *if distances are wrong*.
- Training must be progressive.
- Beware of short cuts. Never plunge in half-way through the stages. Start each exercise from the beginning (Stage 1) until training has become advanced and all the exercises are fully established, with no mistakes being made.
- When in trouble seek advice and assistance.
- When training is interrupted, or things go seriously wrong, it is vital to return to the early stages. Reintroduce the trotting-poles; reduce and simplify the fences.
- Jumping shortened distances will teach a horse to round, bend and use himself, but it is important that the approach should be at the correct pace for his stage of training. Distances that are too long tend to make a horse jump flat (see page 61). Always *teach shortening* before attempting to lengthen.

Schooling Fences, Ditches and Water

When you have mastered gymnastic jumping exercises you can introduce your horse to all the different types of fences including combinations (doubles and trebles), ditches, water and banks, working first in trot and then in canter.

Do not hurry the horse. Give him ample time to jump and to

remain in balance, especially over uprights and parallels. Never use impossible or trick distances (incorrect distances between combination fences).

DITCHES AND WATER

Horses and riders are usually wary of ditches and water. To overcome this problem, careful and progressive training using a confident lead, is necessary. You should start with tiny ditches and work up to larger, deeper ones.

Difficulties arise from:

- Approaching too fast from too far out, thus startling the horse. Always give him time to look—especially in the case of strange or 'starey' obstacles.
- Over-facing him without having built up his confidence by starting with very small ditches.

By looking down into the bottom of a ditch you will cause yourself and the horse to be apprehensive. Always look up and onwards and you will then go onwards.

THE GROUND-LINE *(figure 29)*

A horse assesses an obstacle from the base upwards, and judges his point of take-off from the bottom of the fence, i.e. the most solid part, nearest to the ground and known as the ground-line.

True Ground-line

A clearly defined ground-line will assist the horse's judgement. A pole on the ground pulled out in front of a vertical will help him to arrive at the correct point of take-off. It will become his ground-line—his guide—and he will take-off from it, standing off at the correct distance to jump the fence. For this reason a horse finds it far easier to jump a staircase, or any fence that slopes away from him. He will take off from the bottom—the ground-line—and his natural jump will follow the shape of the obstacle. Therefore no matter how close to the fence he is at take-off, he is unlikely to get underneath it.

False Ground-line

A false ground-line is one in which the base of the fence is set back behind the vertical. The following obstacles all have false

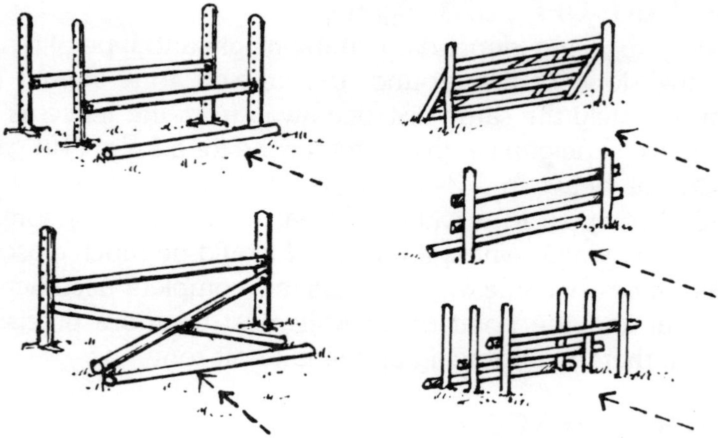

29 **(a)** *True ground-lines.* **(b)** *False ground-lines.*

ground-lines: a gate leaning towards you; upright rails with the bottom pole, or a ground pole, pushed back behind the line of the vertical face; and a triple bar the wrong way on, i.e. with the highest pole nearest to you.

A false ground-line will confuse a horse and cause him to get too close for take-off. He will assume that the top of the fence is directly above the ground-line, and will judge his point of take-off accordingly. Often he will hit the fence by taking off far too close to it.

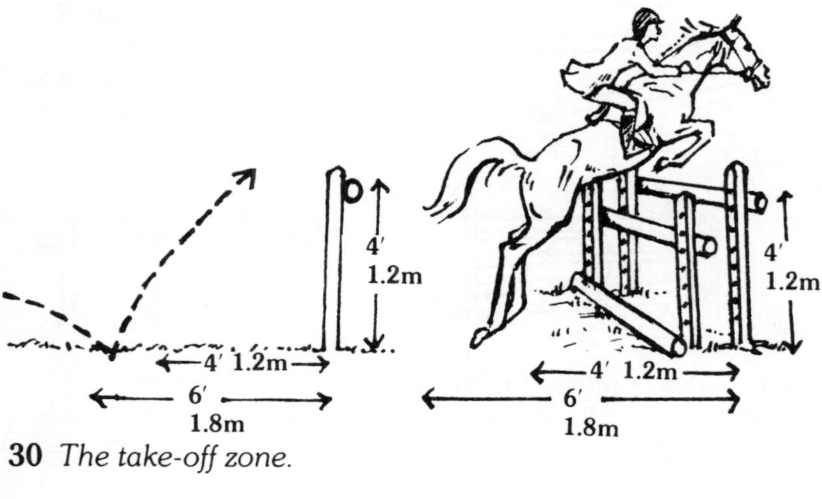

30 *The take-off zone.*

THE TAKE-OFF ZONE *(figure 30)*

Broadly speaking, depending on the height and type of fence and the slope of the ground, the take-off zone should be approximately the same distance away from the fence as its (the fence's) height, i.e. the take-off zone for a 1.2m (4ft) gate should be between 1.2–1.8m (4–6ft) from the base. (This would not apply to a wide triple bar, some types of spread fence or to water, where the take-off would be much closer.)

As fences become larger, however, complete accuracy is essential and the point of take-off must be more precisely judged, thereby narrowing up the take-off zone.

TYPES OF FENCE

There are four basic types of fence, excluding ditches and water. They are listed in order of difficulty, starting with the easiest:

Staircase or Ascending Oxer Triple bars, or a pair of ascending bars.

Pyramid Tiger-trap, hog's-back, etc.

Upright or Vertical Gate, post and rails.

True Parallel or Square Oxer (The most difficult to jump): Parallel bars, planks with a parallel pole behind.

All four types should be included in schooling courses and be incorporated into combination fences (doubles, trebles and quadruples).

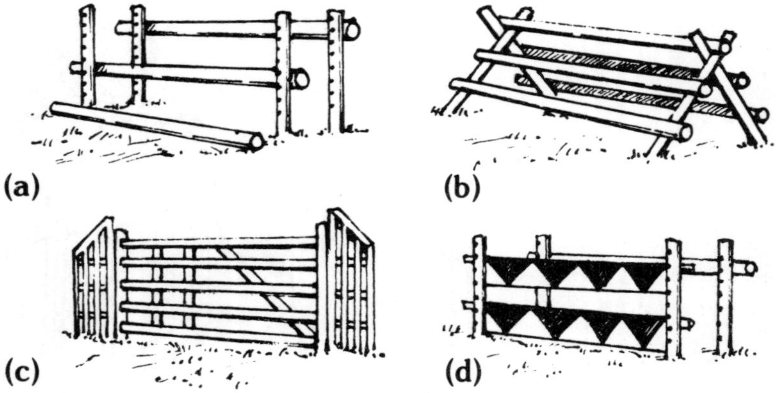

31 *Types of fence.* **(a)** *Staircase, or ascending oxer.* **(b)** *Pyramid.* **(c)** *Upright, or vertical.* **(d)** *True parallel, or square oxer.*

Building Schooling Fences

In schooling much depends on the correct construction of the fences. They should be of inviting, solid appearance, with a good ground-line (see page 80). Poles must be thick and strong—minimum 9cm (3½in) in diameter—and not easily dislodged. Flimsy poles and 'airy' fences do not command respect, and they encourage careless jumping.

Construction of the Ideal Basic Schooling Fence

The staircase type is the most simple and inviting to jump.

Material required: 4 Uprights
4 Poles
4 Cups or pegs
Wings, stands or barrels

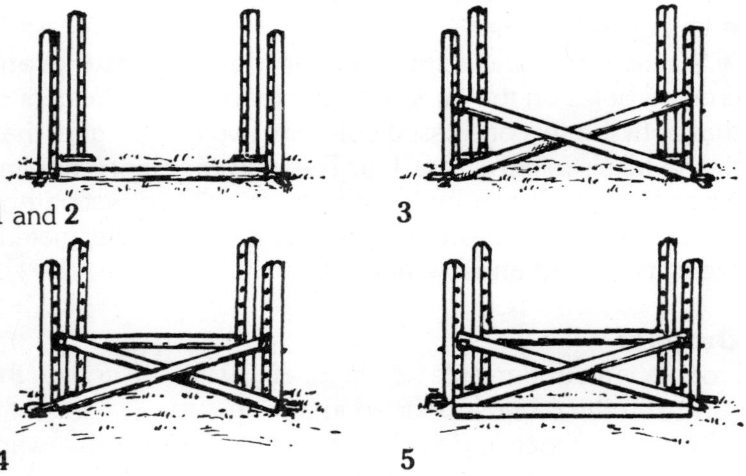

32 *Stages in the construction of a basic schooling fence.*

Method:

1 Set one pole on the ground to show the distance between, and position of, the wings.
2 Place wings in pairs, toe to toe, approximately 0.6m (2ft) apart, at either end of Pole 1.
3 Put two crossed poles on the first pair of wings, approximately 0.7m (2¼ft) high.
4 Add one pole straight across on the second, or back, pair of wings approximately 0.6m (2ft) high.

5 Place a pole on the ground directly in front of the crossed poles (use Pole 1).

- The front pole forms a solid ground-line.
- Crossed poles help to give a more solid appearance. They draw the horse's eye to the centre of the fence and help to keep him straight. He will aim for the lowest part.
- The back pole gives height and calls for more spring and effort than is needed for jumping single or double cross-poles—the very first schooling fences.

With little trouble the basic schooling fence can be changed into:

- Double cross-poles.
- Single cross-poles.
- Ascending oxer (staircase).
- Parallel bar with dropper (see below).
- Upright with variations.
- 'Aachener.' To build this: add one more pair of wings and crossed poles on the far side of the fence; adjust heights so that both the sets of crossed poles are two holes higher than the middle 'straight across' bar. This is an excellent schooling fence because it can be jumped from either direction and has advantages in promoting correct jumping style, making the horse round and use himself.

A Dropper

A dropper is an under-pole set diagonally, with one end on the ground and the other supported in a wing-cup.

It fills up the fence and the lower end can be pulled forward on the 'floor' side to form a helpful ground-line.

JUMPING DISTANCES

The distance between fences depends on the pace of approach, and on the height and type of fence. The measurement is taken from the back of one element to the face of the next element.

Cross-Country Combination Distances

Across country, when galloping on, distances can be longer than those given for approach in trot and canter, and they must be ridden accordingly.

33 *Gymnastic jumping distances with **approach in trot** applying to first two elements only.* **(a)** *One non-jumping stride.* **(b)** *Two non-jumping strides.* **(c)** *Bounce.*

(a)
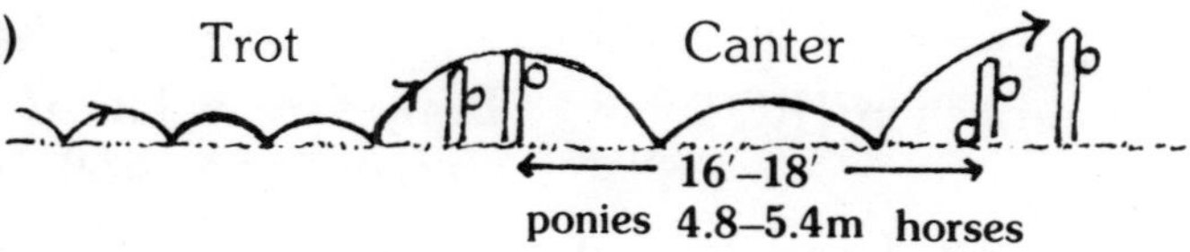

(b)
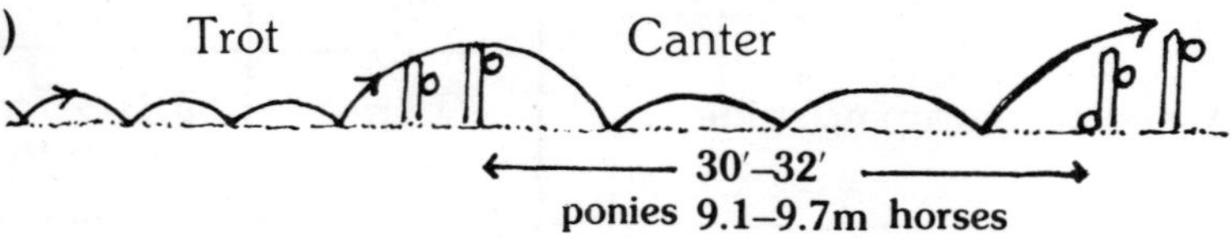

(c)
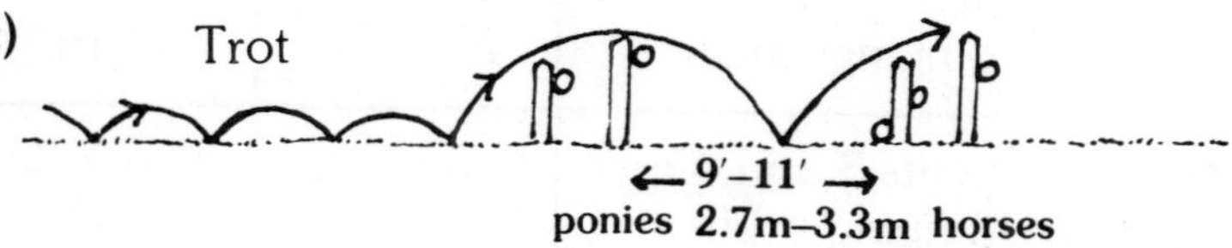

34 *Approximate distances for combination fences with **approach in canter. (a)*** *One non-jumping stride.* **(b)** *Two non-jumping strides.* **(c)** *Bounce. These are 'horse' distances only. For 'pony' distances, refer to Distances for Gymnastic Jumping, page 86.*

(a)
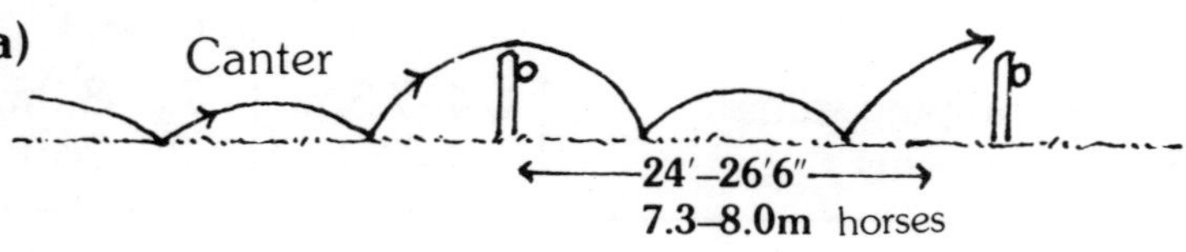

(b)
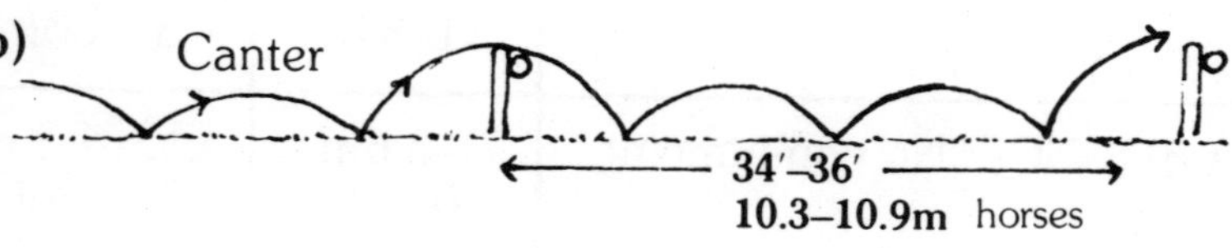

(c)
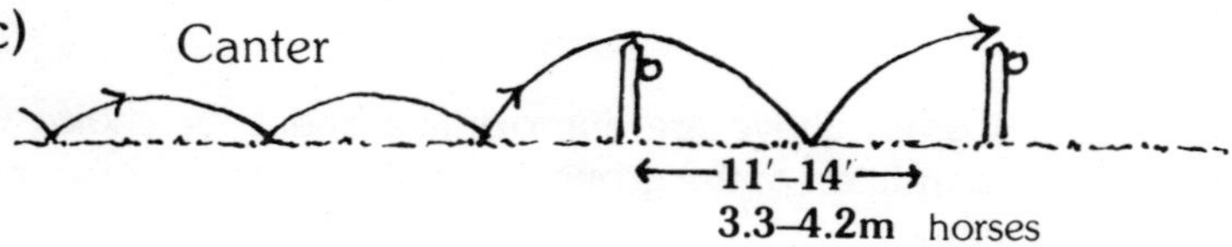

Distances for Gymnastic Jumping* *(figure 33)*
When the approach is in TROT, the distances will be considerably shorter than when in canter and will vary between the different elements.

With the approach in TROT:	Pony (according to size)	Horse
Between elements 1 and 2: for one non-jumping stride	4.8–5.4m 16–18ft	5.4m 18ft
for two non-jumping strides	9.1m 30ft	9.1–9.7m 30–32ft
Between elements 2 and 3: for one non-jumping stride	5.8–6.4m 19–21ft	6.4m 21ft
Between elements 3 and 4: for one non-jumping stride	6.0–6.7m 20—22ft	6.7–7.3m 22–24ft
A BOUNCE: No non-jumping stride	2.7–3.0m 9–10ft	2.7–3.3m 9–11ft

Distances for Combination Fences* *(figure 34)*

With the approach in CANTER:	Pony (according to size)	Horse
One non-jumping stride between elements	6.4–7.3m 21–24ft	7.3–8.0m 24–26½ft
Two non-jumping strides	9.4–10.3 31–34ft	10.3–10.9m 34–36ft
A BOUNCE: No non-jumping stride	3–3.6m 10–12ft	3.3–4.2m 11–14ft

* The distances given above are approximate and your choice will depend on the condition of the ground, the ability of the horse, and the type of fence.

HINTS AND REMINDERS

Hints

- A staircase fence as the first element of a combination encourages novice horses to jump. It is more inviting than an upright.
- A two non-jumping stride double should be used for very novice horses and small ponies. They can put in three strides if necessary.
- Spreads should not be used as a second element for very novice horses or ponies or for those with little scope, especially out of a one non-jumping stride double.
- Never leave empty cups on wings, i.e. spare cups with no poles in them. They are dangerous.
- Use only one pole on the far side of a parallel or near-parallel; you should never have a plank on the far side of this type of fence.
- When making a fence higher, to avoid leaving airy gaps raise all the poles, or put in an extra one.

Reminders

- In early stages of schooling approach in trot.
- Keep fences small until style and complete confidence become established.
- Build inviting fences, that are solid-looking and have a true ground-line.
- Obstacles built beside a school wall or a fence will help novice jumpers—both horse and rider.
- Distances between combinations and in gymnastic jumping must be correct (page 86).
- Parallel bars, double cross-poles and Aachener fences will encourage horses to bend and to round their backs (see Phases of the Jump, Phase 3, page 50).
- Relaxation is vital. Tension ruins style and ability.
- In building jumping courses, other than cross-country, the distance between individual fences should be measured in multiples of 3.5m (4yds), e.g. 14m (16yds), 17.5m (20yds), 21m (24yds) apart, etc. (equivalent to 3, 4 and 5 non-jumping strides).

Unless you are experienced, when setting up or designing a course stick to a simple figure-of-eight track *(figure 35)*.

• For correct schooling it is helpful to know the length of your horse's stride. Approximate lengths are:

	Pony	Horse
Walk stride	0.83m $2\frac{3}{4}$ft	0.91m 3ft
Trot stride	1.0–1.4m $3\frac{1}{4}$–$4\frac{3}{4}$ft	1.3–1.5m $4\frac{1}{2}$–5ft
Canter	depends on size	2.7–3.6m 9–12ft

35 *Jumping courses* **(a)** *Simple figure-of-eight track.* **(b)** *Variations on a figure-of-eight track.*

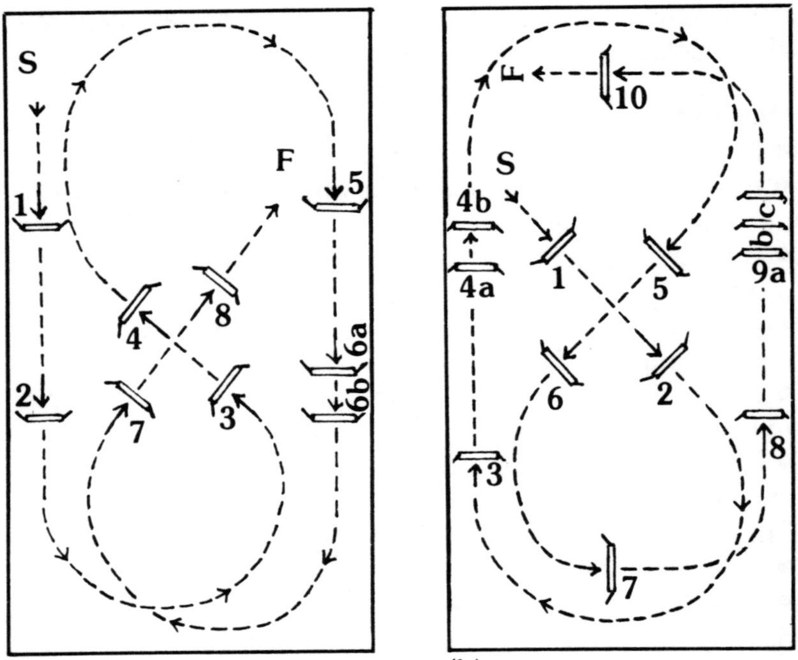

(a) (b)

HORSEMASTERSHIP

Horsemastership
INTRODUCTION

Every good horseman will wish also to be a good horsemaster. To be so he will need the knowledge and experience to deal with all the various problems encountered when looking after a horse. This is of as much importance to the rider who employs a groom as to the owner-groom. To supervise effectively it is essential to understand fully every aspect of horse management—'the master's eye keeps the horses fat'.

The art of horsemastership is acquired by a process of study and experience. Sometimes the experience is painful, as when a kick is suffered or when the loss of a favourite horse results from some error on your part. In dealing with horses there is a right way and a wrong way of doing everything. Sometimes there are several right ways of doing the same thing. In general, the right way is the safe way—safe to the groom and safe to the animal. To teach a right way in all things is the object of all branches of The British Horse Society.

The following pages cover the entire care of horses throughout 365 days of the year, whether the horse is at grass or in the stable, out hunting, or at a Pony Club rally.

1

CARE AND MANAGEMENT OF HORSES

Handling and Leading a Horse

To understand and be in sympathy with a horse's mentality is essential to success. A horse is very much a creature of habit, and favours the same thing being done in the same way at the same time every day. Picking out feet, for example, is much more easily accomplished if they are picked out in the same order each time. Likewise, a horse brought into the stable at night is more easily dealt with if the same timetable is followed each evening. It is a good plan, therefore, to adopt and adhere to, as far as possible, a fixed daily stable routine.

The gregarious instinct in horses is highly developed. They love the company of their own kind. Much that might otherwise prove difficult can often be accomplished easily by keeping stable companions within sight or hearing of one another.

For the rest, the golden rules are: speak quietly; handle gently; avoid sudden movements. There is simply no place in the stable for anyone rough or loud-voiced.

Speaking

Always speak to the horse before approaching; speak before handling; speak before moving. This is a simple but most important lesson to learn.

A word frequently repeated soon becomes familiar. A horse learns much from the inflection of the voice and from the tone and manner in which it is used. He soon comes to recognise the voice of the one who feeds him or from whom some kindness is to be expected. For safety's sake, *speak first*.

The Approach

The approach should always be to the shoulder, taking deliberate steps and never rushing. Speak as you advance and, when you are near enough, pat or stroke the horse's lower neck or shoulder.

FITTING A HEADCOLLAR OR HALTER

Headcollar

This is made of leather or nylon, with a buckle on the near-side cheek-piece. The noseband and throat-lash are connected by a short strap. They may or may not be adjustable—generally they are not. Some headcollars also have a brow band. The headrope is normally attached to a ring in the centre of the back of the noseband.

To fit the headcollar, undo the buckle on the near-side cheek-piece. Place the noseband over the muzzle and the head-piece over the poll behind the ears, then do up the cheek-piece again *(figure 36)*.

36 *Headcollar correctly fitted.*

Halter

This is usually made of webbing or rope. To fit the halter, loosen the noseband and pass the free end of the rope over the neck. Manoeuvre the noseband over the muzzle and the head-piece over the ears. Then tie a knot on the leading side to prevent the noseband from becoming either too loose or too tight *(figure 37)*.

HANDLING THE LEGS

To Lift a Foreleg

First speak to the horse. Then place a hand on his neck and turn to face his tail. Run your hand down over his shoulder, elbow, back of the knees and tendons. On reaching the fetlock, say 'Up' and squeeze the joint. With your free hand catch the toe and hold it with your fingers: less weight falls on the arm when you hold the foot at the toe rather than at the pastern. If the horse won't lift his foot when spoken to, he may be induced to do so if you lean against his shoulder and so push his weight across on to the other leg.

To Lift a Hind Leg

Speak to the horse. Then stand abreast of his hip and face his tail, placing the hand nearest to the horse on his quarters. Run your hand down the back of his leg as far as the point of the

37 *Halter correctly fitted showing detail of knot.*

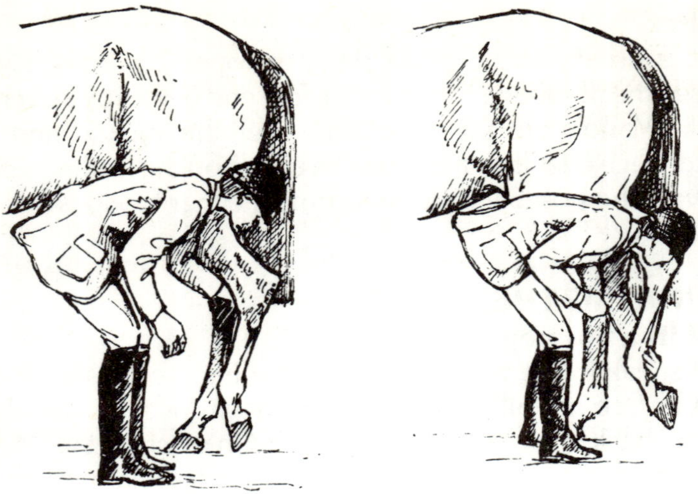

38 *Picking up a hind leg.*

hock. Then move your hand to the front of the hock and run it down on the inside front of the cannon bone. On reaching the fetlock, say 'Up', move the joint slightly backwards when the horse raises his foot, and then slide your hand down to encircle the hoof from the inside. Do not lift it high and do not carry it far back, both of which actions will probably be resisted since they upset a horse's balance *(figure 38)*.

LEADING AND SHOWING A HORSE

To Lead In Hand

A horse should always be led correctly.

A person on the ground leads a horse without a rider by the reins or head-rope. The horse should be trained to lead with his shoulder level with the person leading him.

You should accustom a horse to be led from either side, but since most horses will expect to be led from the near side, with a strange animal you should work from that side. However, in Britain, when leading on the highway the horse must travel in the same direction as the traffic and the person leading him must always be on the off side, between the horse and the traffic (see *The Highway Code*).

It is generally best to lead a horse in a well-fitting snaffle bridle. If the leading area has any access to the public highway a bridle must be worn for control and for insurance purposes.

Horse wearing a bridle Pass the reins over the horse's head and hold them with one hand a short distance from the bit and with one finger dividing them. Hold the buckle end in the other hand.

Horse wearing a headcollar or a halter Hold the rope with one hand a short distance from the headcollar or halter, keeping the free end of the rope in the other hand.

To move the horse Speak to him and walk forward. A trained horse will readily follow. Most horses refuse to move if stared at in the face.

If he hangs back do not pull at his head. Either carry a whip in the outside hand and, moving it behind your back, tap his flank, or, preferably, get an assistant to move behind him.

When turning Steady and balance the horse and then turn him away from you because he then keeps his head up and his hocks under him, thus remaining balanced. He is more under control and therefore less likely to swing out his hindquarters or to kick.

Ride and Lead

Ride and lead is the method by which a mounted rider leads by the reins another horse that has no rider.

A horse should be trained to be led on both sides.

Take the reins of the horse you are leading in the hand nearest to him. With one finger dividing them, hold the reins midway down, keeping your hand close to your knee and as still as possible.

Do not allow the led horse to get his head in front of the one you are riding—nor should you allow him to hang back.

When leading on the public highway you must always keep to the left hand side of the road with the led horse on the near side so that he is sheltered from the traffic.

If either horse is fresh or liable to get upset, keep the led horse on a short rein—about a third of a metre (1ft) from head to hand—with his head about level with your knee. If the led horse tries to break away, give in at first and gradually get him under control, otherwise he may well pull the reins out of your hand or you out of the saddle.

Led horse in snaffle bridle and saddle There are many ways of leading a horse, of which the following are the four most frequently adopted:

- Using the snaffle with the reins straight to the hand.
- With the far rein passed through the ring on the side of the snaffle nearest to you. This prevents a jointed snaffle turning backwards in the horse's mouth if he hangs back.
- With a coupling linking both rings with the lead rope attached.
- With a leading rein joining together the cavesson noseband and the ring of the bit that is nearest to you.

In both of the last two cases the bridle reins must be secured, either round the horse's neck or, if sufficiently long, under the stirrup irons.

The stirrup irons should be 'run up' to prevent them flapping about. Secure them by passing the lower part of the leather through or round the stirrup iron, twice if necessary, or as shown for lungeing (*figure 62*, page 174).

If the leathers fit loosely on the bars you should turn up the safety catches to prevent the leathers slipping off. Make sure that the catches are turned down before the horse is ridden again. Running martingales should be detached from the reins and secured to the neck strap.

Led horse in double bridle Take all the reins over the horse's head and hold them as for the snaffle bridle taking care to see that the curb rein is a little looser than the bridoon rein. Alternatively, isolate and secure the curb rein by crossing the two halves, if they can be unbuckled at the centre, under the neck of the led horse, passing them over and round the neck and re-fastening them under it; then lead with the bridoon rein.

Showing a Horse and Running up in Hand

The object is to show off a horse without a rider either at the halt (showing) or while moving as freely as possible at the walk and trot (running up in hand). The general procedure is the same as for leading in hand.

Showing Make sure that the horse is looking alert and standing still so as to show his conformation (see page 192) to the best of advantage. Stand directly in front of and facing him, holding one rein in each hand. Raise your elbows so that the horse cannot nibble your wrists. If the horse is wearing a headcollar or a halter, place your hands on each side of the noseband with the end of the rope in one hand.

Running up in hand When leading past for inspection, place yourself on the far side of the horse so that you are not between the horse and the person inspecting him.

When running up away from or towards the person inspecting, it is usual to place yourself on the near side. Run or walk as instructed, looking straight ahead and not at the horse—he should be moved directly away from and directly towards the person inspecting who will step to one side if he wishes you to take the horse past him.

It is most important that the rein or rope should be slack enough to allow the horse to carry his head naturally. He should be encouraged to move confidently and well. The pace should be active, but not hurried or unbalanced.

The Stabled Horse

It should be the ambition of every good horseman to strive for the highest attainable standard in the turn-out of his horse, in the management of the stable, in the care of tack, and in the cleanliness of the premises. Cleanliness is particularly important. Germs, disease and infection thrive in dirty conditions.

CONSTRUCTION AND EQUIPMENT OF STABLES

Loose Boxes and Stalls

Loose boxes These provide the most satisfactory stable accommodation. The horse has more freedom of movement, is encouraged to lie down and rest, and is consequently more comfortable than in a stall. To accommodate a hunter, a box needs to measure about 4.3m (14ft) by 3.6m (12ft); to accommodate a pony, about 3.6m (12ft) by 3.0m (10ft).

Access to a loose box is either by way of a stable door direct from the yard or by a stable door and passage-way within the stable. In the former case the door ought to be in two parts so that the upper portion can be hooked back and left open allowing the horse to look out over the lower portion. Horses

(a) (b)

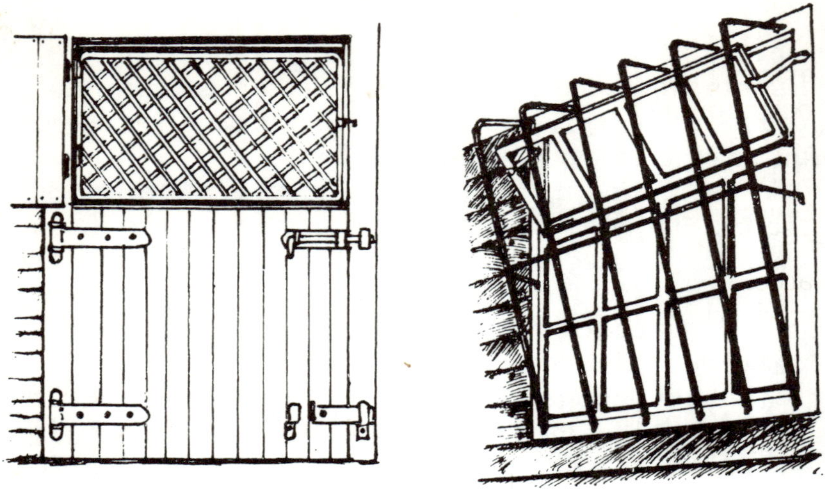

39(a) *Stable door with grille and two bolts, the lower one foot-operated.* **(b)** *Ventilation window hinged to open inwards and with protective bars.*

greatly appreciate the chance to see what is taking place in their immediate neighbourhood. Grilles are sometimes fastened across the open, top half of the door to prevent the horse jumping out, biting at passers by, loosening the top bolt or 'weaving' *(figure 39a)*.

Doors should be at least 1.25m (4ft) wide. Narrow doorways are dangerous. It is important that the doors of loose boxes should open outwards so that you can enter the box without interference from bedding. This is also important if an animal should become cast in his box near to the door.

Latches should be of a special non-projecting type so that there is no risk of the horse being injured by them when passing through the doorway. Two are needed, one at the top of the door and another at the bottom. The lower one prevents damage through kicking and it is an advantage if it is of the foot-operated type. An overhang or verandah outside a row of boxes is useful since it provides protection from the weather.

Stalls These have the advantage of providing accommodation for a greater number of animals in a given area and of saving labour and bedding material.

Disadvantages are that the horses are unable to move around and may become bored because they cannot look out; nor do they have direct access to fresh air. A shy or nervous animal may be deprived of rest as a result of bullying by his neighbour in the adjoining stall.

Access to stalls is normally by way of a passage within the stable.

Swinging bail stalls *(figure 40)* These are sometimes used in Pony Club camps and by riding schools for the temporary accommodation of horses caught up from grass. They are easily cleaned, permit free circulation of air and facilitate drying-off of the standings. The bail should be approximately the length of the stall. It must be hung high enough to prevent a horse getting a leg over it, and be slightly higher off the ground at the front end than at the back. It should be suspended by a stout cord tied with a quick-release knot; should a horse get a leg over the bail by any mischance, he can then be released quickly. Chains or wire should not be used. One of the great disadvantages of the swinging bail is that it provides no protection against bullying and neck-biting.

40 *The use of a swinging bail as a stall division.*

Floors

It is essential that surfaces be non-slippery, impervious to moisture, and long-wearing. A variety of materials is available. Concrete is cheap and easily installed. If used, be sure that it is given a roughened facing. On chalky soil which drains well it may not be necessary to have an artificial floor. This makes for quiet, and is also restful to tired legs and joints.

Stable floors must slope slightly to allow for drainage, but no more than the minimum amount for the purpose. Drains should be either in the corner of the box away from the manger, haynet and door or outside the box or stall. Gullies leading towards them should be shallow and open. If shavings or peat are used for bedding, care should be taken to see that drains are kept clear, or that the opening is covered up.

Ventilation

This is an important matter. Given enough fresh air, horses keep healthier and are less prone to coughs and colds. Warmth should be provided by excluding draughts and by means of extra clothing rather than by restricting fresh air. The simplest plan is to leave the top of the stable door open, day and night, summer and winter. This is almost always possible, provided that the stable faces south and that the bottom portion of the door is of sufficient height to exclude draughts. For this reason a stable with a south-facing aspect is recommended. There should also be a window, preferably on the same wall as the door to prevent through draughts, which should be glazed with reinforced glass or have the glass protected by iron bars. It may be combined with a window hinged so that it opens inwards from the top allowing the incoming air to circulate upwards and over, not on to, the back of the horse *(figure 39b)*. For greater protection from draughts, partitions between boxes and stalls should extend right up to the roof.

Electric Light Switches

These ought always to be outside the stable, never inside, and they should be in a position where the horse cannot interfere with them. They should be of a special stable type designed to prevent electrocution should a shod horse seize hold of one with his teeth.

Fittings

The more free of encumbrances a box or stall is, the better. Fittings should, therefore, be kept to the minimum.

Securing rings In a stall the horse is tied up all the time, and must be able to lie down without getting his legs entangled in his headrope, so two rings are essential:

- One fixed to the wall at a height of approximately 1.5m (5ft) is for short-racking and for the haynet. When used for racking up, the rope must be passed through a small loop of string attached to the ring.
- One fixed either to, or alongside, the manger (if fitted) is for normal tying up. This should only be used in conjunction with a log attached to the free end of the rope (see *figure 41).*

A haynet ring fitted independently of the short-rack ring and at about the same height, is convenient but not essential.

In a loose box, where a horse is unlikely to be tied up for any length of time, only the short-rack ring is essential.

All rings must be really firmly fixed.

Mangers Placed at approximately the same height as the horse's chest and either in a corner or along a wall, a manger needs to be shallow enough to prevent the horse getting his jaw caught in it, deep enough to prevent him throwing his feed out, and preferably with the rim broad enough to prevent him biting it. Removable plastic mangers which either hook over the door or are fixed to brackets on the wall are convenient and easy to keep clean. Improvised mangers are rarely a success. Fitted mangers may include a separate compartment for hay.

Hay racks fitted above head level These are now generally condemned, as the horse has to feed at an unnatural height with the risk of dust and hay seeds falling into his eyes.

Water bowls Gravity-fed or lever-operated bowls fixed to the wall, each with a separate stop-cock, are favoured by some people. They should not be sited near the manger or haynet in case they become blocked with food. They must be cleaned daily. In some stables, large hinged rings are fixed to the wall about 1.2m (4ft) off the ground to hold the feed or water bucket so that it cannot be knocked over. When not in use, they can be folded down against the wall.

When fitted mangers, hay-racks and water bowls are dispensed with altogether, it is easier to clean the stable, and feeding is carried out at ground level—a much more natural position—using a movable metal, plastic or heavy rubber feed tin or a wooden feed box. Circular tins, about 46cm (18in) in diameter and wooden boxes with sloping sides about 38cm (15in) high and 46cm (18in) square at the top, are long-lasting and easily cleaned. Hay is fed loose on the ground or, more economically, from a haynet secured to the short-rack ring.

Utensils

The following are essential:

For feeding A plastic bowl or bucket in which to mix the feed and take it to the manger, feed-box or feed-tin; a haynet; stable buckets.

For mucking out A wheelbarrow; a stable shovel; a broom; a pitchfork and four-pronged fork, each with blunted prongs; a skep for droppings; a hose-pipe; a sheet of material such as sackcloth or hessian for carrying straw; a rake, if peat or shavings are used.

Utensils should be stored neatly, outside the box or stall, preferably under cover, or in a nearby building, so that the risk of injury to animals is avoided. Ideally, all but the barrow should be hung on hooks on the wall.

SECURING A HORSE IN THE STABLE

Horses should be secured by a headcollar and head rope, the latter always being attached to the back 'O' ring of the nose-band.

To Secure the Horse by 'Log' and Rope

The head rope is passed through a manger-ring and secured to a 'log' in such a way that the 'log' just touches the ground as the horse stands up at the manger. The 'log' takes the slack out of the rope, thus preventing the horse getting a leg over it and becoming cast *(figure 41)*.

To Short-Rack (Rack up)

Tie the horse with a short rope to a ring on the wall at a height of approximately 1.5m (5ft). This should be done whenever it is necessary to restrict the horse's movements, to prevent him

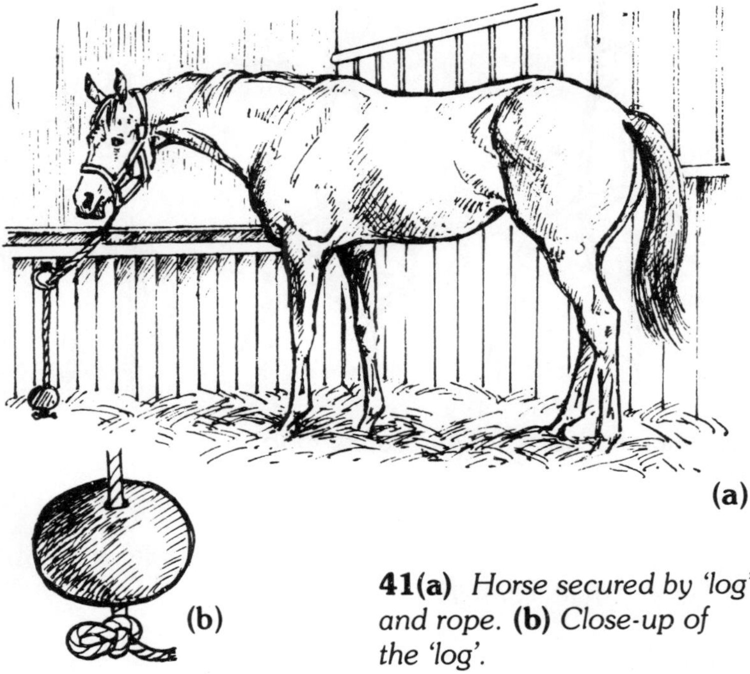

41(a) *Horse secured by 'log' and rope.* **(b)** *Close-up of the 'log'.*

from turning round and biting during grooming or from lying down after being groomed.

To Place a Horse on Pillar Reins

Turn him round in his stall, fasten the reins or chains (if fitted) which hang on the heel-posts, to the side 'D's of his headcollar. Pillar reins have various uses, in particular to keep a saddled or bridled horse under control until he is required.

When tying up a horse, it is most important that some recognised form of quick-release knot should be employed. The securing rope should be passed through a small loop of string (ordinary string, not nylon) attached to the ring rather than through the ring itself. This will secure the horse for normal purposes, but will give way should he panic and struggle hard to free himself; otherwise, should his headcollar break, or the ring be pulled out, he is very likely to go over

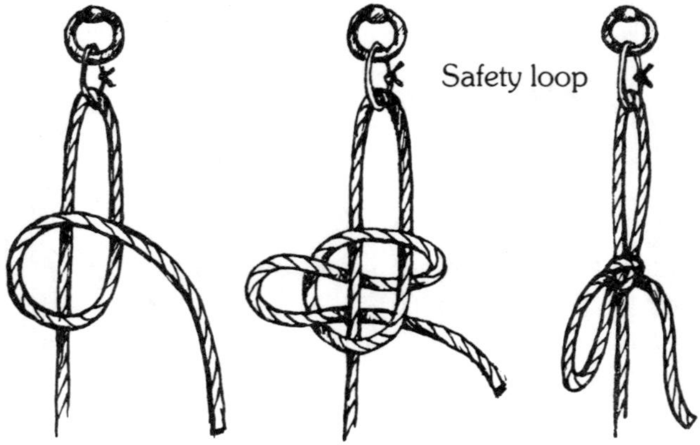

42 *Quick-release knot and safety loop.*

backwards and damage himself. A simple and efficient form of knot is shown in *figure 42*.

BEDDING

Some form of bedding material is necessary for the stabled horse, to allow him to lie down and rest, to encourage him to stale, to provide insulation, and to prevent his feet being jarred during long hours spent standing on a hard surface.

Straw

Wheat straw is, on the whole, the best type of straw for bedding. It is normally easy to obtain. It is also convenient to store when baled; it gives a bright appearance to the stable; it makes a warm, comfortable bed; it is easily handled; and it permits free drainage. *Oat straw* may also be used as bedding; but, because it is more palatable, especially if from a crop sown in the spring or undersown with grass, horses are inclined to eat it. Also, being more porous it quickly gets wet and soggy. *Barley straw* is more satisfactory than oat straw and makes good bedding provided that it comes from a combine harvester that has removed the awns, which are prickly and irritate the skin. It is often longer and of better texture than 'combined' wheat straw, and is usually less expensive. However, horses do have a habit of eating it.

Management of straw bedding First thing in the morning muck out the box or stall:

- Pick up all droppings and separate the soiled portions of the bedding from the clean with a pitchfork and discard only the dung and wet portions.
- Throw the dry bedding back to the sides of the stall or box and put the soiled portion on sacking or in a wheelbarrow placed at the entrance. This allows cleaning, airing and drying of the floor.
- After exercise, replace bedding so that the horse can lie down to rest and also to stale, which few horses will do on a bare floor because of the risk of slipping and their dislike of splashing their legs.

In the evening, re-make the bed adding fresh straw as necessary. The floor should be swept clean and scrubbed out with disinfectant every week or two.

When bedding down, the used straw should be put down first and fresh, clean straw added; or, if the horse tends to eat his bedding, it is better to mix the old and new straw together. Baled straw should be well tossed when making the bed. If it is trussed, the straws should lie across and not parallel to each other. As an extra precaution against injury or draughts it is usual to bed more thickly around the walls of a loose box or the sides of a stall. A good, deep bed makes for comfort and is more economical in the long run.

If the horse eats an excessive amount of straw bedding you should either spray the straw with disinfectant or replace it with shavings or peat. Alternatively you can muzzle the horse, but this is not really satisfactory except for a limited period such as the night before hunting.

Droppings should be removed each time you visit the stable. This is done by lifting the straw from beneath and tipping the droppings into a basket or tin, called a dung skep. A plastic laundry basket is a satisfactory substitute.

Peat Moss

This is popular and makes an excellent bed. It is particularly valuable where there may be a risk of fire. Wet and soiled patches must be changed frequently and the bed forked and raked over daily. This keeps the peat soft and sweet and prevents it from becoming packed and soggy. As with straw, droppings should be removed each time you visit the stable. When first put down, peat moss is apt to be a somewhat dusty bedding. The disposal of soiled peat is sometimes a problem but when rotted it is very good for gardens, especially for those on heavy soil.

Shavings and Sawdust

Used either separately or together, shavings and sawdust make a good, clean, bright and comfortable bed that is often also economical. When used together, the sawdust should form the under-layer. As bedding, shavings and sawdust require the same attention as peat.

Before shavings, sawdust or peat are used, drains, traps, etc., must be carefully sealed off.

Deep Litter

The method of bedding known as the 'deep litter system' is used by many horse owners. Under this system, which uses either straw, shavings or peat, droppings must be removed at every opportunity and fresh bedding added to an existing deep bed when necessary. The advantages are a slight economy of bedding, a saving of labour—except when periodically the whole is cleared out—and the provision of a deep, warm bed which does not need to be shaken up each day. With adequate ventilation, enough fresh bedding, and the frequent removal of droppings, there should be no offensive smell.

Depending on the height of the stable, the whole bed should be removed and re-started when it becomes more than about a third of a metre (1ft) deep—say, every three to four months. Unless you have plenty of assistance or can use a tractor with a fork-lift attachment, this is a very heavy and lengthy task. When peat or shavings are used, the damp patches can be removed periodically and the bed thus kept at a constant depth.

Disposal of Manure

'To judge a man's stable management, visit the manure heap.' Slack, lazy methods will be reflected in a pile of only partly-expended straw. An owner or groom who handles his bedding correctly is likely to be careful and energetic in his other duties.

If the manure is kept in a pit near the stables, arrangements should be made for the pit to be emptied very frequently or for the manure to be sold. If the manure is to be stacked, choose a place some way from the stables and store it in three heaps: the oldest pile consisting of well-rotted manure ready for use on the garden; a second pile to which manure is no longer added, being in the process of rotting; and a third pile in current use as a dump for fresh manure.

A PROGRAMME OF STABLE ROUTINE

The following specimen programme is intended to serve as a guide to the manner in which the various stable duties may be fitted into the day's work and to show the hour at which each task is best peformed. It allows for three or four feeds in the course of the day.

Considerable adjustment will, of course, be necessary to meet varying conditions and individual cases.

The care of a horse on a hunting or competition day, or on the day of a Pony Club rally is dealt with on pages 177 to 185.

7.00 am **Stables:**
Look for the 'signs of good health' in your horse (see page 261).
Put on headcollar and tie up.
Look round him to see that he has suffered no injury during the night.
Adjust rugs.
Clean water bucket, re-fill and replace.
Tie up small net of hay.
Muck out.
Pick out feet.
Throw up rugs and quarter.
Replace rugs.
Remove headcollar.

8.00 am **First feed** (A small feed if the horse is to be exercised within an hour-and-a-half.)

9.00 am Put on headcollar and tie up.
Remove droppings.
Remove rugs.
Saddle up.
Exercise.

	On return from exercise wash off and pick out feet; remove saddle and bridle.
	If bedded on straw, spread enough to encourage the horse to stale, roll and have a drink.
	Put on headcollar and tie up; sponge out eyes, nose and dock.
	Set fair; rug up and remove headcollar.
11.30 am	Tie horse up and give him a small net of hay.
	Groom.
	Put on day rug(s).
	Re-fill water bucket.
	Remove headcollar.
	Second feed.
	Tie up a net of hay.
	Set fair stable and yard.
2.00 pm	Set fair stable, clean tack, attend to grass-kept horses and general maintenance.
4.30 pm	Put on headcollar and tie up.
	Remove droppings.
	Pick out feet.
	Shake up bedding.
	Remove day rug.
	Rug up with night rug.
	Fill water bucket.
	Remove headcollar.
	Tie up a net of hay.
	Third feed.
7–9.00 pm	Set fair. Remove droppings; re-fill water bucket; shake up bedding.
	Fourth feed (if necessary).
	Re-fill haynet.

The Grass-Kept Horse

Advantages

Far less attention is required under this natural system. If it is properly managed many horses thrive on it and there are considerable savings in time and money.

The grass-kept horse exercises himself sufficiently for health but not for fitness. To keep fit and 'hard' he needs the same amount of exercise as the stabled horse. Because he is not restricted, the grass-kept horse is less likely to be over-fresh when required for work.

Grass, the natural food, becomes the mainstay of the diet while there is feed in it—normally from May to September—but it must be supplemented during the winter months and during the summer if the horse is working. Provided that he is getting the exercise appropriate to the work he is being asked to do, a grass-kept horse is less likely to suffer injury to wind and limb than a stabled horse, since his legs, and to some extent his lungs, are being exercised constantly.

Disadvantages

Horses are not always near to hand when required and may be very dirty and wet. The field may be some distance from the house, or the horse may refuse to be caught. If land is not available at home, there may be real difficulty in finding a field which can be rented or a farmer who is willing and able to accept a horse for grazing.

GENERAL MANAGEMENT

A horse turned out to grass should be visited at least once a day and looked over for injuries, condition of feet and/or shoes, and general indications of health (see page 261)—head alert, eyes wide open, ears pricking to and fro, moving sound with strides of even length and, when still, with the weight distributed evenly over all four feet or resting a hind, but not a fore, foot. The coat should be lying smoothly and in warm weather it should be glossy. If it is dull and standing up (i.e. 'staring') this indicates poor condition or cold, sometimes both. If the former, seek expert advice; if the latter, put on a New Zealand rug.

A horse being worked from grass should also be visited at least once a day and looked over even when he is not being ridden. It is as well to catch him every day in order to ensure that this can be easily done when he is required for work. A few oats, pony cubes, slices of apple, etc, in a feed tin, bucket or scoop will bring the horse to you. Approach the horse quietly towards the shoulder, or let him come to you, and avoid jerky or sudden movements that might frighten him off. Mud should be removed before work; sweat marks, when they have dried out, should be removed after work and his eyes, nose and dock sponged out. Avoid excessive grooming which removes the grease from the coat, nature's protection against wet and cold. An unclipped horse is unlikely to have dried off completely on return from work, but is less likely to catch a chill if turned straight out than if left to stand in the stable to dry off.

The Seasons

Late summer An abnormally dry summer may result in grassland becoming more bare than at any other period of the year. In extreme cases, the grass may become so brown or burnt up that no keep for the horse remains. When this happens the horse must either be moved to where better keep is available, or fed hay as a temporary measure. It must not be forgotten that in a dry summer the normal water supply may also fail, in which case you must take water to the horse by other means.

Mid-winter In the winter months the time arrives when the feeding of hay, as a supplement to grass, becomes necessary. This ought not to be put off for too long or the good condition carried by the horse from the summer will be lost. A thin horse is also a cold horse. In cold weather horses tend to spend their time sheltering rather than grazing. At such times additional hay should be provided and even a feed of concentrates—oats, cubes, etc, (see Feeding). Feed should be provided in that part of the field in which the horse shelters. Snow, while causing no acute discomfort to him, naturally puts a stop to grazing, and the quantity of hay should then be increased.

The early months of the year, when the coat is at its longest, are also the months of lice infestation. This should be looked for and the necessary action taken to control it (see page 299).

Early spring Hay and concentrates should be continued as supplements until there is real feed in the grass—this is often not until late May.

Early summer Some horses and many ponies, particularly small ponies, tend to grow too fat during the summer months. This should be discouraged since a fat pony is never a pleasant ride, and its excessive weight is dangerous because an attack of laminitis may result (see page 285). A horse or pony that shows a tendency to put on too much weight should be moved to a field offering bare keep only, so that he gets less to eat and has to move further in search of it.

Flies are a great trial at this time of year. If possible the horse should be provided with a shelter shed. Failing that, shade and the companionship of another horse with whom he can stand head to tail so that each can flick away the flies from the other's face, is appreciated. There are also a number of insect repellants on the market. However, flies could well be nature's way of keeping a horse at grass active and on the move rather than just standing about and filling himself up with grass.

ROLLING

The grass-kept horse has a great advantage over his stabled companion in that he is able to indulge in a roll in natural conditions whenever he feels so inclined.

Why do horses roll? The reason is by no means clear. Rolling is undoubtedly done in spring-time to help shift the winter coat and also frequently after work, the inference being that those parts of the skin subjected to pressure need a rub. The object may also be to cover wet portions of the coat with dust so as to assist the drying-off process; a layer of mud helps to keep out the cold and wind, and to retain warmth. It is obvious that rolling is also an act of sheer enjoyment, and is to be encouraged as natural, healthy and relaxing.

GRAZING LAND

Acreages

It is impossible to lay down any hard and fast rule about the acreage needed to support a horse, as so much depends upon the nature of the soil, quality of the grass, drainage of the land, etc. But if the horse is at grass throughout the year, one acre per horse may be taken as a minimum and two acres per horse as generous, provided the land is in good order.

Conservation

The keeping of a horse at grass requires careful and intelligent use of land. Horses are among the most wasteful of grazing animals. In their search for the most palatable grasses they trample down and destroy potentially valuable food. Furthermore, in the course of time, portions of the field become so stained with droppings that the grass becomes rank and no longer acceptable as food. If the land is to be used economically it is essential that a long-term plan be adopted. The following are some of the ways in which waste may be checked or avoided.

The daily removal of droppings by shovel and barrow is, perhaps, the most effective of all conservation methods. The benefits are undeniable. Failing this, periodic harrowing to spread dung works well, and the horse himself may be expected to help by drawing a brush-harrow.

If horses are allowed unrestricted access to all the grassland available, then by mid-June little of the grass growth associated with that time of year will be in evidence.

By dividing a piece of land into three parts with fences and grazing each part in turn, two portions can be set aside to recover or be rehabilitated while the third portion is in use. This method has the added advantage of allowing one part to be laid up from April to June each year for the taking of a hay crop, thus easing the problem of winter keep. Under this plan you will need to provide for three sources of water and shelter. Alternatively, you can confine your horse to one part of the paddock from late April until the hay has been made, supplementing his grass feed from existing stocks of hay whenever this becomes necessary.

The grazing of the land for a period of time by other types of stock such as bullocks or sheep is also an excellent plan. They eat off grass rejected by horses and also reduce worm parasites in the grass which are dangerous to horses but harmless to themselves. If other stock are not available, frequent cutting of the ungrazed growth will make it more palatable and so increase your grazing area.

FENCES

The fencing of land for use by horses differs from that needed by other stock in two essentials. Firstly, the risk of injury to a horse is greater, and secondly, there is the possibility that a horse will jump the fence.

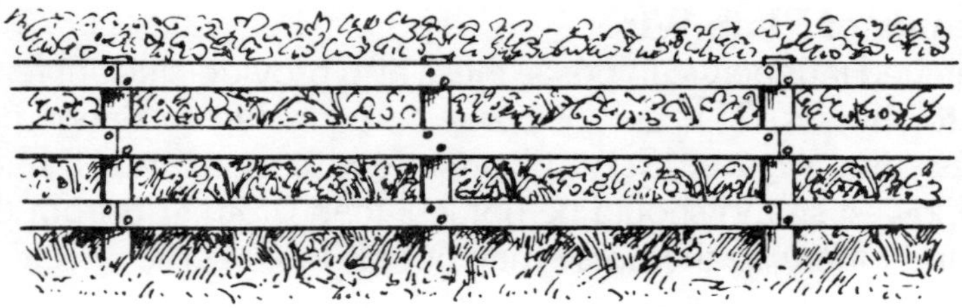

(a)

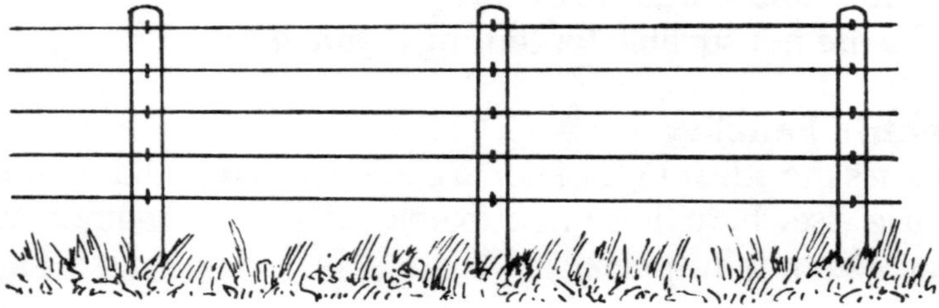

(b)

(c)

43 *Types of fencing.* **(a)** *Post and rail within a shelter hedge.* **(b)** *Post and wire; the bottom strand must be at least 0.3m* **(1ft)** *from the ground.* **(c)** *The type of fencing NOT suitable for horses.*

Post and rails

These are the first choice especially when positioned inside the existing hedging. However, the expense of erecting them and the cost of their maintenance are high *(figure 43a)*.

Hedges

If tough, strong, and well-cared for, hedges are the second choice and have the additional advantage of providing wind-breaks and shelter in all weathers. With the exception of yew and deadly nightshade, the hedges of the countryside contain nothing poisonous to horses.

Strands of Plain Wire

Stretched taut between posts, these also provide efficient fencing for horses. The use of barbed wire is highly dangerous unless taut, *(figure 43c)* and preferably backed with timber. The lowest strand should be not less than 0.3m (1ft) from the ground otherwise the horse may put his foot over it.

Hurdles and Chestnut Paling

These are not suitable for fencing *(figure 43c)*.

Electric Fencing

This has the advantage of being easily erected and dismantled for use elsewhere. It is undoubtedly an efficient method, and horses learn to respect it.

Shelters

Wind-breaks Rain may cause horses to stop grazing, but otherwise it worries them little. Snow, too, is a matter of little consequence. It is for shelter from wind that provision ought to be made. Wind-breaks or wind-screens formed by thick coppices, or even a high hedge along the north side of the field or along the side from which the prevailing wind blows, offer excellent protection and horses generally prefer them to a shelter shed.

Shelter sheds The horse can use these at any time and they provide protection against all the elements. Experience shows,

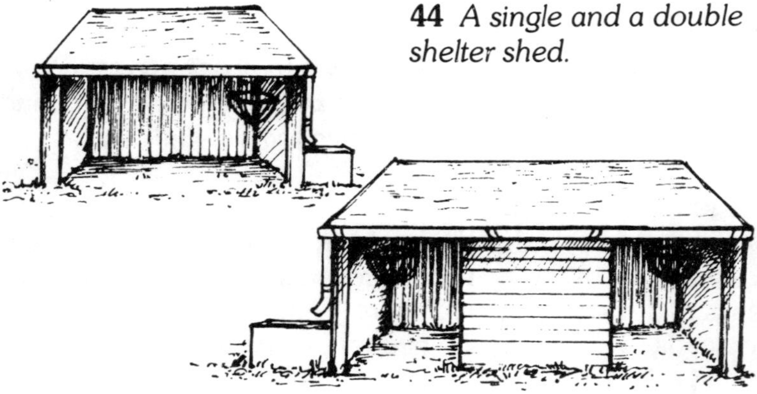

44 *A single and a double shelter shed.*

however, that horses seldom resort to them in winter; they really come into their own during the summer when horses greatly appreciate the protection that they provide against flies, and on this account their use is fully justified. Ideally, shelter sheds should be placed in a corner of the field, back to the prevailing wind, and be easily accessible for feeding. If the shed is built circular in shape or with a double doorway it is less likely that when two horses are turned out together one will become cornered and be injured by the other. Cobwebs should not be removed from a shelter shed as they act as a useful trap for flies *(figure 44)*.

The Combined System

This is a system of stable routine which allows the stabled horse to spend a portion of each day at grass. It is a compromise between stabling and grass keep and, by allowing use to be made of both methods, has much to recommend it.

It is a system well suited to an animal required frequently for showing, hunting or other purposes, where labour is scarce and time for regular daily exercise not available. The rule in such cases is to run the horse at grass by day and to stable him at night, except during the heat of the summer when the procedure is reversed. This arrangement, besides ensuring that the horse is in the stable when required, provides proper facilities for grooming and corn feeding.

It is under this system that the New Zealand rug comes into its own since it overcomes the problem of turning out a clipped animal in winter time (see page 116).

Clothing

RUGS, BLANKETS, ROLLERS AND ROLLER PADS

The following are the articles of horse clothing commonly used.

Rugs and Blankets

These are used to provide warmth in the stable by day and by night. They may be of either man-made or natural fibre.

Man-made fibre is non-absorbent and so, unlike natural fibre, does not 'breathe' thus tending to cause the horse to break out into a sweat in mild weather or, after exercise, before he has fully settled. Suitably lined, man-made fibre rugs are light and easy to wash, but they also tear easily.

Day rugs When of woollen material and bound with braid of another colour, they are excellent and give the best appearance. Hemp, jute, or man-made fibre rugs, suitably lined, are equally efficient as day or night rugs.

Night rugs Made of hemp, jute or man-made fibre suitably lined, they often have a webbing surcingle sewn on to them. This may lead to pressure on the spine and it is advisable to cut it off and use it with a pad or a roller instead. Night rugs are likely to get soiled when the horse lies down so it is as well to have another set for use by day to keep the horse's appearance smart.

Horse blankets Made of wool or man-made fibre, they are used on a clipped horse under rugs during the colder months of the winter.

Rollers, Roller pads and Breast-girths

Rollers of leather or webbing Used to keep rugs in place. They must be correctly fitted, with a clear passage between the padding where it rests on either side of the spine.

Roller pads Used under rollers to prevent pressure on the spine. They must be thick, and wider than the padding on the roller.

Breast-girths Used to prevent the rollers slipping back. They should be attached to the 'D's on the front of the roller.

New Zealand Rugs *(figure 45)*

These rugs are traditionally of waterproof canvas partially lined with wool, and are designed for use on a horse at grass. The canvas type is heavy. The man-made fibre types are much lighter but may not be so durable, and are more expensive. New Zealand rugs are not intended for wear in the stable but are valuable as a rug on a horse turned out to grass in cold weather, especially if clipped or partially clipped. Their special merit is that they provide protection against wind and rain and do not become dislodged when the horse lies down and rolls.

45 *A New Zealand rug. The leg-straps are crossed over between the hind legs.*

Some New Zealand rugs have a surcingle sewn on to them, the ends passing through the sides of the rug. Other types have no surcingle and thus avoid the risk of causing a sore back. Leg-straps are provided to prevent the back of the rug folding over. To avoid chafing, these should be crossed over between the hind legs before being passed around the hind thigh.

Projecting parts of the horse such as withers must be checked daily for signs of chafing. It may be necessary to sew sheepskin pads into the rug's lining to protect these parts. A horse needs to become accustomed to a New Zealand rug before being turned loose in it, as the stiffness of the material and the rub of the leg-straps may upset a nervous horse the first time he wears it.

Rugs and Sheets for Use in Special Circumstances

Anti-sweat rugs Made of open cotton mesh, these are popular for use as coolers for an over-heated horse. They may be used on their own or in place of straw under a natural—not man-made—fibre rug on a horse that is sweating or tending to 'break out' on return from work.

Waterproof rugs Made of various materials, these are useful at competitions or other outdoor events in wet weather.

Summer sheets Made of cotton, these are used to protect the groomed horse against dust and the worry of flies. They ought to be provided with a surcingle and a fillet string, which crosses behind the quarters to prevent the sheet from blowing about in the wind.

Putting on a Rug

The following procedure is the most practical:

- First, gather up the rug or sheet and throw it well forward over the horse's back.
- Sort out the front part and fasten the buckle.
- From a position level with the horse's hip pull the rug back into place, using both hands.
- Check and, if necessary, adjust the position of the rug or sheet. If the horse is absolutely trustworthy this can be done from immediately behind. This is one of the few occasions when it is permissible to stand close behind a horse. If the horse is not reliable, adjust the rug from the side and check it from further back, out of reach of his heels.
- Place the roller in position with the buckle on the near side of the horse.
- Place the roller pad, if used, under the roller.
- Move round to the off side and see that the roller hangs well forward, i.e. where the girth goes, otherwise it will slip forward and be loose when the horse moves.
- Buckle up the roller on the near side, firmly but not too tightly.
- Smooth down the rug underneath the roller on both sides by running your fingers down between roller and rug, at the same time giving the rug a slight pull to ease it forward in front of each elbow, thus preventing drag on the points of the shoulders.
- Do up the breast-girth, which, if used, should lie loosely above the points of the shoulders.
- Finally, check the pressure of the rug around the neck and shoulders. It is important that the rug should fit the horse, i.e. that it is neither too large nor too small. In either case, apart from other discomfort caused, sore withers may result.

If a blanket is used as well as a rug, then it should be put on first and in a similar manner. Care should be taken that an ample portion lies on the neck and that the back of the blanket does not extend beyond the root of the tail. The rug is then added and, after adjustment of the roller or surcingle, the surplus portion of the blanket lying on the neck is folded back as an extra precaution against slipping.

Taking off a Rug

From the near side of the horse proceed as follows:

- Unbuckle the roller or surcingle and pad, and the breast-girth, if used.
- Remove the roller and, after folding it up, place in a corner of the box.
- Unfasten the breast-buckle of the rug and, using both hands, fold back the front portion of the rug (and blanket, if used) over the top of the back portion.
- With your left hand on the centre front and your right hand on the centre back, remove the rug together with the blanket, if used, in one gentle backward sweep following the direction of the lay of the horse's coat.

Alternatively, rugs may be taken off one at a time in this manner and either folded up immediately or, with the roller or surcingle and pad, placed temporarily over the door. The blanket or rug worn nearest to the body should be shaken vigorously outside the stable.

When not in use, all rugs and blankets should be folded neatly and placed in the corner of the box or on top of the roller or surcingle and pad. Now and again, on a dry day, hang them outside to air and freshen up.

BANDAGES

It is important that before use all bandages should be correctly rolled. The tapes, unless replaced by clips or Velcro, should be neatly folded across the width of the material which is then rolled with the sewn side of the tape inwards.

Stable Bandages *(figure 46a)*

Made of wool or stockinette, 10cm (4in) wide and supplied in sets of four, they provide warmth and prevent legs that have got wet while working from chapping. In skilful hands they are very beneficial, but if they are wrongly or carelessly applied

(a) *Stable bandage*

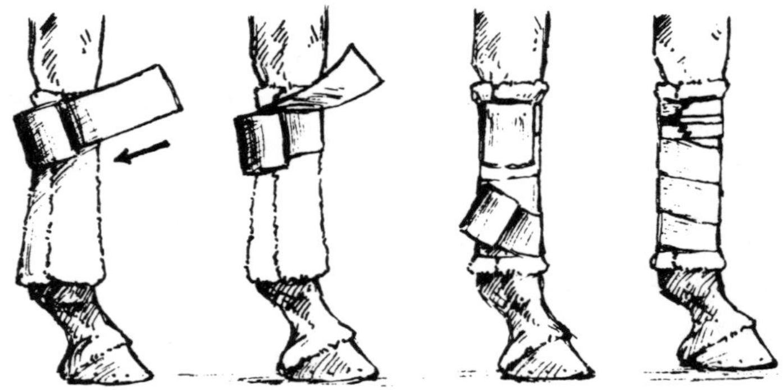

(b) *Exercise bandage*

46(a) *A stable bandage should cover the leg from the knee or hock to the coronet.* **(b)** *An exercise bandage should finish above the fetlock so that the bandage does not interfere with its action.*

they become a danger. Ringed marks on a leg or a lump on a tendon indicate a bandage put on carelessly or too tightly. The use of gamgee or softened straw under a bandage guards against this.

To bandage a leg Start just below the knee or hock, passing the bandage round the leg in even turns until the coronet is reached. Here the bandage will take a natural turn upwards. Continue unrolling in an upward direction and finish off at the place where you started. Fasten the tapes in a bow and tuck in the spare ends *(figure 46)* or secure by clip or Velcro. The tapes should be tied so that the knot or bow lies to the outside or the inside of the leg, and not in front where it will press on the bone, nor at the back where it will press on the tendons.

Exercise Bandages *(figure 46b)*

Used primarily to protect the legs while the horse is at exercise or at work. They are approximately 7cm (3in) wide and are made of stockinette, crêpe or various synthetic materials, which should be slightly elastic. They should cover the leg between the knee or hock and the fetlock, but should not interfere with the movement of either joint.

Apply the bandage over gamgee or some other type of padding such as cotton wool or foam rubber. First wrap the gamgee (or its substitute) around the leg, taking care that it lies flat. Start just below the knee (or hock) at the side of the leg, and bandage in the same direction as the overlap of the gamgee, from top to bottom and up again, making the cross-over just above the fetlock joint, in the centre at the front. Each turn around the leg should cover about two-thirds of the width of the bandage. The pressure should be even, and sufficient to ensure that the bandage does not slip, but not so tight as to cut off the circulation or to impede movement of the tendons.

Finish off by folding over the beginning of the bandage and covering it as the top is reached. This fold should not lie across the back of the leg. The tapes should lie flat and be of the same tightness as the bandage. For general use, tie them in a knotted bow, on the outside of the leg, and tuck the ends in carefully. For special occasions, or for galloping, use a reef knot. For extra security, cover with plastic insulating tape or sew the ends of both tapes and bandages.

The protruding gamgee may be trimmed with scissors.

Tail Bandages

These should be 6.5–7cm ($2\frac{1}{2}$–3in) wide and made of stockinette, crêpe or synthetic material. They are used to save the tail from injury or rubbing when travelling, to improve the appearance of the tail, and to keep the hairs of a pulled tail in position.

Tail bandages should not be used to excess. They should never be left on all night, since if you want to prevent them from falling off and getting tangled in the legs or the straw you will almost certainly put them on too tight, which stops the

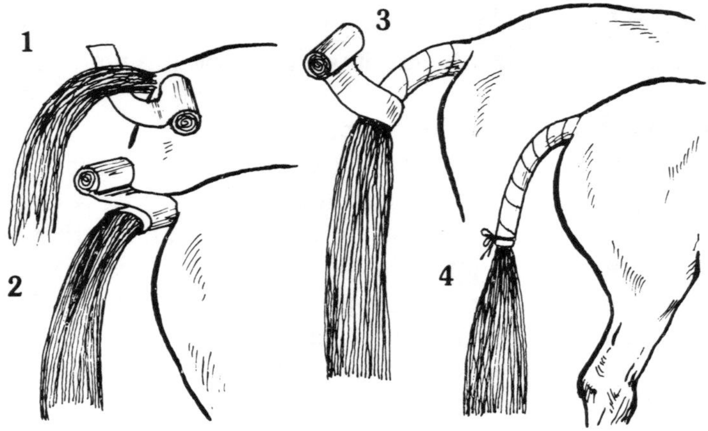

47 *The correct way to bandage a tail.*

horse's circulation and damages his tail. In the case of a stabled horse, tail bandages should be put on after exercise and grooming and removed at evening stables. When hunting, etc., they should be put on first thing in the morning and removed before hacking on or after unboxing.

To apply the bandage, first slightly dampen the tail hair with a water brush. Wetting the bandage is wrong as this may shrink the material and injure the tail, and it will certainly cause the hairs of the tail to lie in the direction of the roll. Place your left hand under the tail, unroll about eight inches of bandage and place this spare piece under the tail, holding its end in your left hand and the roll of the bandage in your right. Keep your left hand on the root of the tail until the spare end is secured. The first turn is often difficult to make securely but you can overcome this by making the next turn above the first. Unroll the bandage evenly around and downwards stopping just short of the last tail-bone. Tie the tapes neatly, but no tighter than is necessary to secure firmness. Then bend the tail back into a comfortable position *(figure 47)*.

Removing Bandages

To remove a bandage from a leg Untie the tapes and unwind the bandage quickly, passing the freed parts from hand to hand. Do not try to roll up the bandage while you are removing it. When you have removed it, give the back tendons and the fetlock a brisk rub with the palms of your hands. Then hang the bandage up in order to air and dry.

When applying bandages it is dangerous to kneel in the vicinity of a horse's legs. You should adopt a bending or crouching position.

To remove a bandage from the tail Untie the tapes, grasp the bandage with both hands at the point nearest to the dock and slide it off the tail in a downwards direction.

BOOTS

There are numerous types of boot designed to prevent a horse injuring himself or being injured. The injuries involved are described on pages 281 to 282.

The principal types, each produced in a variety of materials and designs *(figure 48)*, are:

Brushing These include fetlock boots, Yorkshire boots and rubber rings.

Speedicut Similar to brushing boots but fitted higher on the leg.

Coronet To prevent 'treads' by other horses, principally when playing polo or travelling. They are normally made of thick padding with a leather covering.

Knee See Travelling (below).

Hock See Travelling (below).

Over-reach Bell-shaped and made of rubber, they fit round the lower portion of the pastern.

Tendon To provide protection to the tendons against high over-reach. They have a strong pad at the rear, shaped to the leg. They are sometimes shaped to guard also against brushing.

Travelling To prevent injury while on a journey. Types include: *Travelling boots* which encase the legs completely from above the knee or hock down to and overlapping, the coronet. *Hock boots*, normally made of thick felt and leather. *Knee caps* which may also be worn when exercising on the road or when schooling for

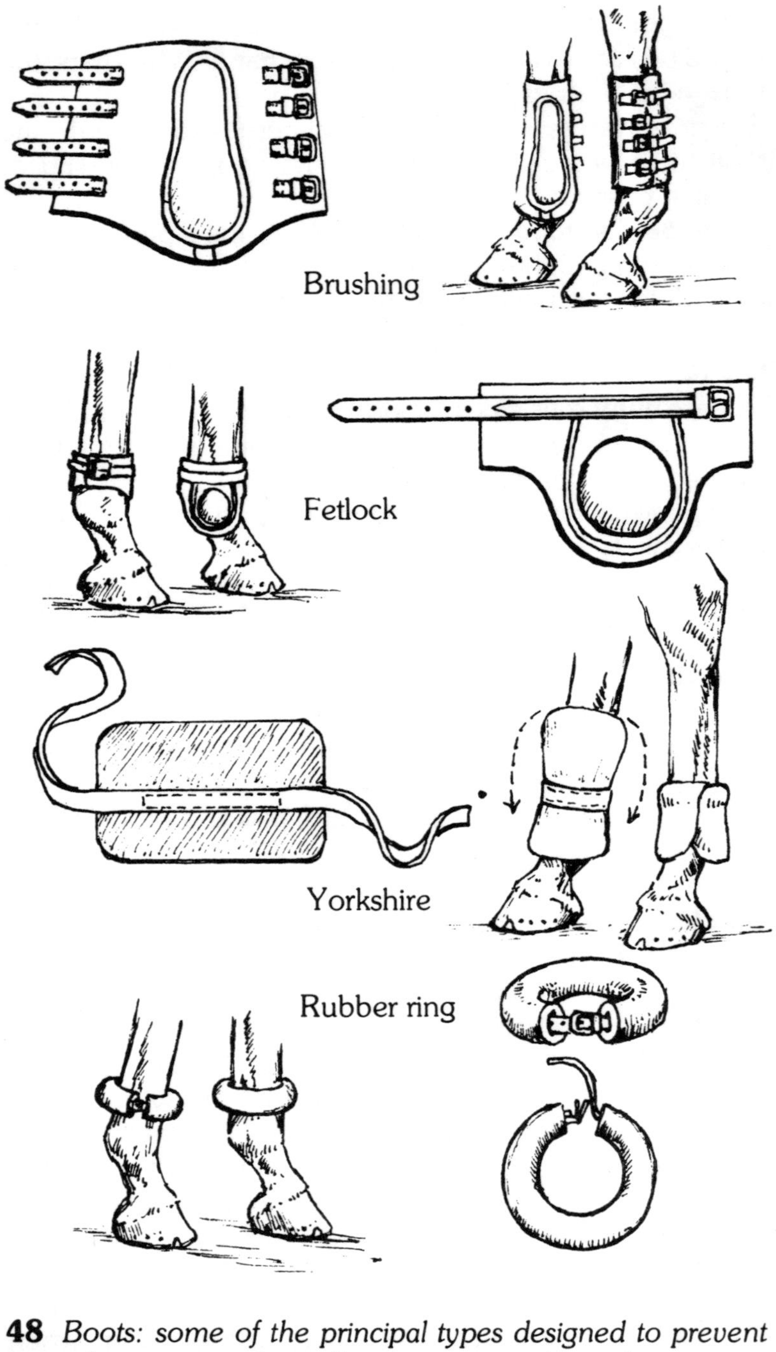

48 *Boots: some of the principal types designed to prevent a horse from injuring himself or being injured by others.*

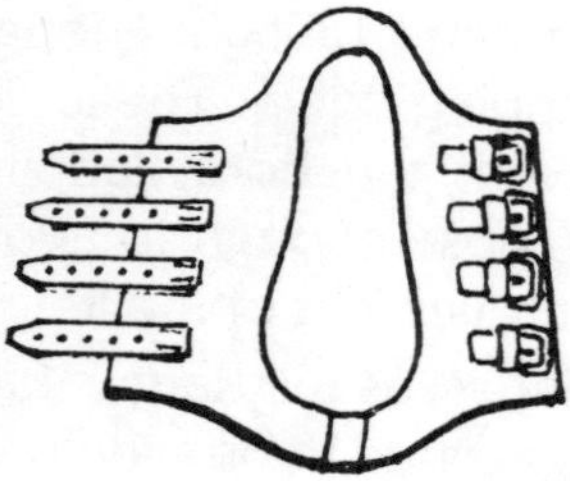

Speedicut

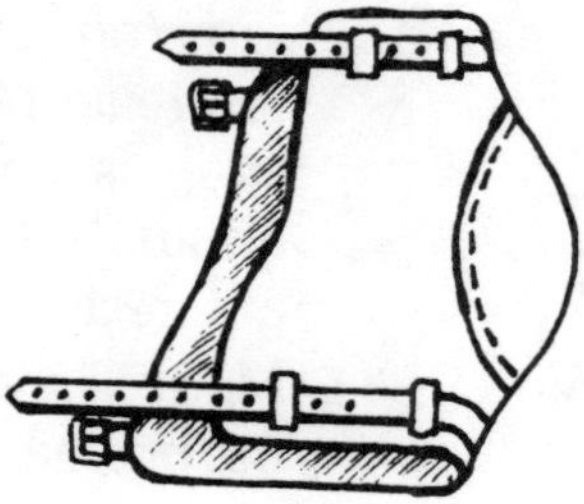

Hock

Over-reach

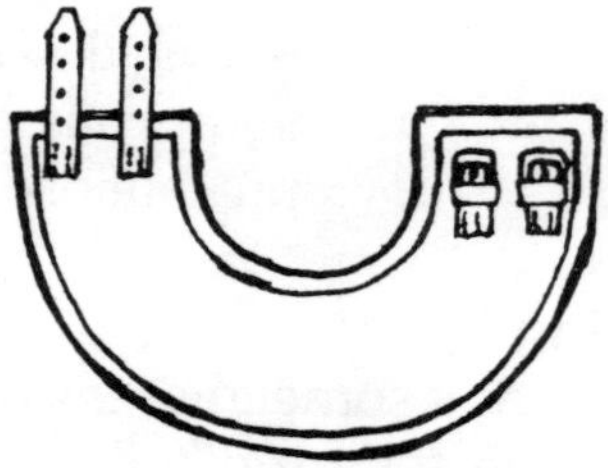

Coronet

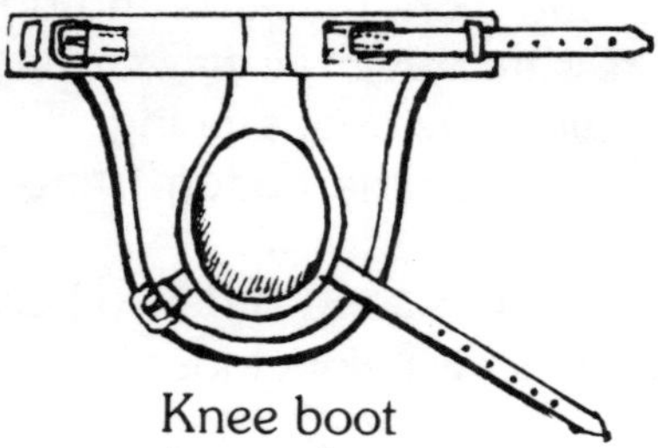

Knee boot
(or cap)

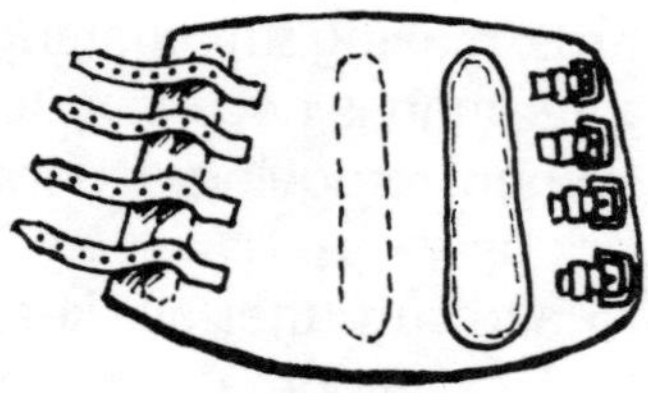

Tendon

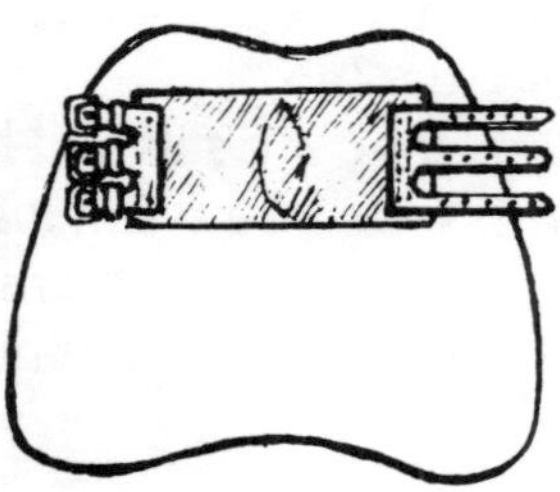

Polo

jumping. The top strap must be tight enough to prevent the boot slipping down. The lower strap is there to prevent the boot flapping about, but it must be loose enough to avoid any restriction of the movement of the leg.

Polo Made in a variety of shapes, usually larger than the normal boot and designed to provide protection against blows from polo balls or sticks, brushing, speedicut, 'treads', etc.

Sausage Fitted round the coronet to prevent the heel of the shoe damaging the elbow when the horse is lying down. They are only required when an injury has been sustained, not as a normal precaution.

Fitting

When, as is sometimes the case, the front and hind leg boots of the same set are different in shape, make sure that you put them on the right way round.

Straps should be on the outside of the leg with the ends pointing towards the rear of the horse.

Start with the middle strap and work upwards and downwards, making sure that the pressure is even all the way and no tighter than is necessary to prevent the boot slipping.

When removing a boot, start from the bottom strap and work upwards.

Make sure that all 'keepers' are in good order, otherwise buckles can come undone when straps work loose.

Care

Boots must be thoroughly cleaned after use, as any mud, etc, left on them will harden and may cause sores later on.

Depending on the material of which they are made, clean them as described for saddlery (see page 255).

Keep any straps as soft as possible. If they are stiff and hard, apart from the danger of cracks and breaks you will have the greatest difficulty in doing them up and in undoing them.

Care must be taken with boots that have metal clips which, if knocked, may come unfastened.

GENERAL CARE OF HORSE CLOTHING

Clothing is an expensive item of equipment, but with proper care it can be made to last for years. Regular brushing and an occasional airing or sunning when not in use is desirable.

In summer, winter clothing should be washed—scrubbed in the case of jute or canvas—and, when dry, folded and stored away with moth-balls, since moth is one of the commonest causes of deterioration to horse clothing. Alternatively, send winter clothing to a cleaner's or laundry before storing.

Grooming

Grooming is the daily attention necessary to the coat, the skin, the mane, the tail and the feet of the stabled horse. In the case of the grass-kept horse some modification is required although the importance of a methodical approach remains the same. For notes on grooming the grass-kept horse see page 134.

An experienced groom will spend from half to three-quarters of an hour on the task. A novice will probably need longer because of fatigue resulting from the unaccustomed use of certain muscles. Thoroughness, however, brings its own reward, for few things are more pleasing to the eye than a well-groomed horse.

Grooming kit *(figure 49)*

Body brush For the removal of dust and scurf from the coat, mane and tail.

Curry comb Made of metal for cleaning the body brush, or of plastic or rubber for removing caked mud from a horse kept at grass.

Dandy brush For removing heavy dirt, caked mud and dust. It is of special value for use on the grass-kept horse.

Hoof pick For picking out the feet.

Hoof oil For oiling the hooves.

Mane comb

Stable sponges For cleaning eyes, nose, muzzle and dock.

Stable rubber For a final polish after grooming.

Wisp (or massage pad) For promoting circulation, and for massage.

Water brush For use, dampened, on mane, tail and feet.

HOW TO GROOM

The following description of the procedure to follow, although by no means an exhaustive account, covers all essential points.

After Exercise

Although it is desirable to keep a horse's feet free of mud, it is by no means essential to wash them every day; in cold weather it is better not to do so. When you do wash them, use the water brush dipped in a bucket of water. The thumb of the hand which holds the horse's foot should be pressed well into the hollow of the heel to prevent water being lodged there.

In Stable

Collect the articles of grooming kit. They are best kept together in a box, wire basket, or even a bag.

- Put the headcollar on the horse; then tie him on the short rack, and remove his rugs. If the weather is cold, leave one blanket folded over his loins while grooming his front, and over his shoulder while grooming his quarters. It is often more pleasant in summer to go outside and find a suitable spot to tie him up.

- Begin by picking out the feet with the *hoof pick*. Pick up each foot in turn; remove whatever may be lodged in it with the point of the pick, working downwards from the heel towards the toe. Working this way there is no risk of the pick penetrating the soft parts of the frog. Clear the cleft of the frog and look for any signs of thrush (see page 287). Tap the shoe to see that it is secure and, finally, run the tips of the fingers round the clenches to see that none is risen.

  When picking out the feet it is permissible to lift the off feet from the near side, a practice to which a horse soon becomes accustomed.

  A dung skep should be placed near enough to allow dirt to fall directly into it, and so prevent it dropping on to the bed or having to sweep it up afterwards.

- Now take the *dandy brush* and start at the poll on the near side. The object is to shift all caked dirt, sweat marks, etc. Certain parts of the body come in for special attention—e.g. the saddle region, head, points of the hocks, fetlocks and

49 *Grooming kit: the items are described on page 127.*

pasterns. The dandy brush may be held in either hand and is used with a to-and-fro motion. It may be helpful to grasp the tail with your free hand when working on the hind limbs. The use of the dandy brush on the tender parts of the body, and on most clipped horses, is best avoided; a rubber curry comb used in the same way is an excellent alternative and is also quite effective for cleaning the body brush.

• Next take the *body brush.* The object and also the procedure are now quite different. The short, close-set hairs of the body brush are designed to reach right through the coat to the skin beneath, and the brush must be used in such a way that this is achieved *(figure 51)*.

□ Begin with the mane. First, throw the mane across to the far side of the neck and thoroughly brush the crest. Replace the mane and then start work on it, beginning at the head end. Insert a finger of your free hand into the mane so as to separate a few locks of hair. First the ends and then the roots are brushed to remove tangles. Work slowly down the neck, dealing with only a few locks of hair at a time.

□ Pass to the grooming of the body. Begin at the poll region on the near side. Take the body brush in your left hand and the metal curry comb in your right. For the most part it is easier to have the body brush in your left hand when working on the near side, except for those parts under the stomach. Stand well back, work with a slightly bent arm and a supple wrist and lean the weight of your body behind the brush. Use short, circular strokes in the direction of the lay of the coat rather than to-and-fro. After every four or five strokes, draw the brush smartly across the teeth of the curry comb to dislodge the dirt. The curry comb in its turn is cleaned by tapping it out on the stable floor, or in the passage, and not against the wall or manger. When the near side has been completed, pass to the off side, change hands and repeat the process.

□ Now do the head. First untie the horse, turn him towards the light and remove the headcollar. If the door of the box is closed, there is no need to secure the horse. If it is open, fasten the head-strap temporarily round the neck. Put down the curry comb and use your free hand to steady the head.

Use the *body brush* carefully to avoid injuring tender parts or knocking bony projections. When you have finished, replace the headcollar and tie the horse up again.

• Now take the *wisp*. Wisping is a form of massage to develop and harden muscles, to produce a shine on the coat by squeezing out the oil from the glands in the skin, and to stimulate the skin by improving the blood supply. Dampen the wisp slightly and use vigorously by bringing it down with a bang in the direction of the lay of the coat. Give special attention to those parts where the muscles are hard and flat, such as the sides of the neck, the quarters and thighs. Avoid, however, all bony projections and the tender loin region.

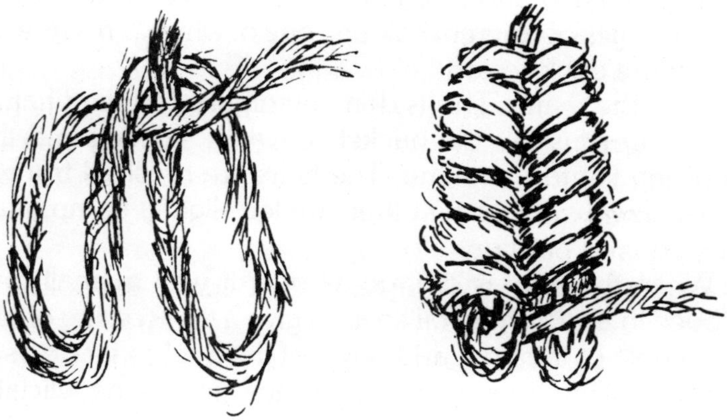

50 *How to make a wisp.*

□ *To make a wisp* Make a tightly woven rope about 1.8 to 2.5m (6 to 8ft) long by twisting up hay or straw *(figure 50)*. Soft hay is best and should be dampened slightly first. Fashion two loops at one end of the rope, one slightly longer than the other. Twist each loop in turn beneath the remainder of the rope until all is used up. The far end of the rope should then be twisted through the end of each loop and finally tucked away securely under the last twist. The hay wisp should then be dampened and stamped on. A properly made wisp should be hard, firm and small enough to be easily held in the hand. Alternatively, leather-covered felt massage pads are obtainable and can be used for the same purpose.

- The *sponge and bucket of water* are required next. 'Sponging out' refreshes a stabled horse and is appreciated perhaps more than any other part of the grooming routine.

Wring out the sponge so that it is soft, clean and damp. Start with the eyes. Sponge away from the corners and around the eyelids. Wring out the sponge again and deal with the muzzle region, including lips and the inside and outside of the nostrils in that order. Wring out again. If the horse is reliable, move behind him to attend to the dock. If he is a kicker or is not reliable, stand at one side to do this. The tail is lifted as high as possible and the whole dock region, including the skin of the under-surface of the tail, is gently sponged and cleaned. Some people prefer to keep two sponges of different colours, one of which is reserved for use in the dock region.

- 'Lay' the mane. This is done by dipping the end hairs of the *water brush* in the bucket of water, shaking and then applying flat to the mane. The hairs are brushed from the roots downwards so that they are left slightly damp and in the required position.
- When the hoof is dry, go all over it with a small brush dipped in a jar of *hoof oil* so as to give a thin coating of oil to the whole of the hoof and bulbs of the heel as far up as the coronet. This improves appearance and is beneficial to broken or brittle feet.
- Go all over the horse with the stable rubber to remove the last traces of dust from the coat. For this purpose the rubber is made up into a flat bundle, dampened, and the coat wiped in the direction of the lay of the hair.
- Finally, the tail is dealt with. Take only a few strands of hair at a time. Do this by holding the tail and shaking a few strands free, brushing the ends clean first. The use of the dandy brush on the mane or tail is wrong as it removes and breaks the hairs, leaving them thin or unsightly.
- When grooming has been completed, put on a tail bandage, laying the top hairs first with a water brush.

Tail Washing

Required: a bucket of warm water; a cake of mild soap or a little suitable shampoo (detergents and soft soap should not be used) and a clean body brush. The procedure is as follows: first soak the tail in warm water to remove all dirt; then shampoo and rinse thoroughly, changing the water as necessary; remove the tangles.

If the horse is quiet, the end of the tail may with advantage be immersed in the bucket. Squeeze out the water with your hands and then swing the tail to dislodge any water remaining. Do not bend the hairs. Next, using a vertical movement, brush out the tail, a few hairs at a time, with a clean body brush. This will clear any remaining scurf. Finally, apply a tail bandage.

WHEN TO GROOM

Quartering

This is done first thing in the morning to remove stable stains. The object is to make the horse look tidy before exercise and to give those parts of the coat which have to be dampened to remove stains a chance to dry off before the horse goes out. The feet are picked out. The eyes, nostrils and dock are sponged. Rugs are then thrown up or unbuckled in front and turned back. The parts of the body exposed receive a quick brush down. Particular attention should be paid to the removal from the flanks of stains acquired when lying down at night, the sponge or the water brush being used for this purpose. Note that the whole operation is carried out without undoing or removing the roller or surcingle. Remember to rebuckle the front strap of the rug.

Strapping

This is best carried out on return from exercise. It consists of the entire grooming procedure given on pages 128-132, which is always more effective after a horse has been exercised since exercise warms up the skin, loosens and raises the scurf to the surface and opens up the pores. If, however, a horse is fed on return from exercise, a short time should be allowed for him to take his feed undisturbed.

Brush Over or Set Fair

In the evening the horse is given a light brush over and wisping when the rugs are being changed, and the box set fair, i.e. droppings removed and bedding tidied up.

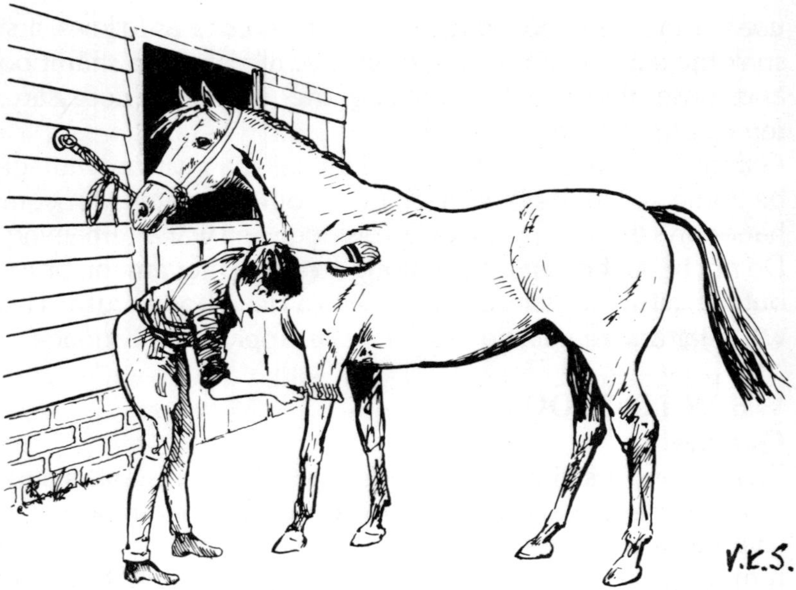

51 *The correct use of the body brush (see page 130).*

To prevent the littering of the stable paths or yard, feet should be picked out several times a day, and always when a horse leaves his box or stall.

Grooming Machines

Provided they are used intelligently and the maker's instructions are followed, grooming machines are a boon in a large stable. Great care must be taken over their introduction and subsequent use.

GROOMING THE GRASS-KEPT HORSE

If a horse is living a natural life the skin will be in a thoroughly healthy condition and it is unnecessary to groom him unless he is being ridden. Under these conditions grooming should be limited to attention to the feet; a brush down with the dandy brush or rubber curry comb to remove mud and sweat marks; use of the body brush to keep the mane and tail tidy; and the sponging out of eyes, nose, muzzle, and dock.

The efficient use of the body brush other than on the mane and tail is scarcely possible on an animal that rolls every day, nor can it be used to any effect on a horse with a long winter coat. In any case, grease and dandruff should not be removed from the coat of a horse running out, as they contribute both to body warmth and to the 'waterproofing' of the hair. For further information see the Pony Club publication *Keeping a Pony at Grass*.

Clipping and Trimming

CLIPPING

The reasons for clipping a horse include the following:

- To enable a horse to carry out fast work without undue distress.
- To conserve condition by avoiding heavy sweating.
- To permit a horse to work longer, faster and better.
- To facilitate quicker drying off on return from work.
- To save labour in grooming.
- To prevent disease.

Types of Clip

The full clip The whole of the coat is removed.

The hunter clip *(figure 52a).* As above, except that the hair is left on the legs as far as the elbows and thighs, and a saddle-patch is left on the back. For this system it is claimed, with considerable justification, that the coat left on the legs acts as a protection against cold, mud, cracked heels and injury from thorns, and that the saddle-patch saves a sore or 'scalded' back under the saddle. Legs may be carefully trimmed (see How to Clip, page 136).

Many hunters are clipped right out the first time, and the legs and saddle-patch left at the second clipping. On a common horse this has the advantage of making the legs appear less hairy than would otherwise be the case.

Great care must be exercised in carrying out this clip neatly, particular attention being paid to the saddle-patch. If it is too far forward the horse will look short in the shoulder and long in the back. However, if it is cut straight behind the shoulder and allowed to come slightly back behind the saddle it will often greatly improve the appearance of the animal.

The blanket clip The hair is removed only from the neck and belly, a patch corresponding in size to that of a blanket being left on the body.

The trace-high clip *(figure 52b)*. The hair is removed from the belly, shoulders and thighs up to the level at which the traces would run, and is left on the legs as for the hunter clip. Sometimes the hair immediately under the neck is removed also. This clip is useful in the case of horses kept at grass and is a compromise between clipping fully and not clipping at all. It is a common form of clip for harness horses.

The belly and neck clip The hair is removed from under the belly upwards between the forelegs, and up the lower line of the neck to the lower jaw.

Equipment

Clipping machines are of three types: hand clippers, wheel machine clippers, and electric clippers. *Hand clippers* are slow and laborious, *wheel machine clippers* are out-dated and rarely seen nowadays. *Electric clippers*, although expensive, are by far the most satisfactory type.

Before starting, it is important to ensure that the machine is well oiled and the blades sharp. A horse is easily upset by a noisy motor or by hot and pulling blades; periodically, therefore, in the course of the clip the head of the clipper should be stripped and thoroughly cleaned, oiled, and allowed to cool. When using the type of electric clipper which has the motor enclosed in the body of the handle, care must be taken to keep the air filter clear, or the motor will over-heat.

How to Clip

Under any system clipping is a lengthy process, so enough time must be set aside for the task and the necessary assistance arranged.

The coat must be dry and as well-groomed as its length will permit—dirt clogs and over-heats the clippers. Tie up the tail with a tail bandage to keep it out of the way. As much of the horse as possible should be clipped without upsetting him or resorting to any means of restraint. The start may be made anywhere, but in a nervous horse begin at the shoulder region. The most difficult parts to clip, both from the point of view of

(a)

(b)

52(a) *A hunter clip.* **(b)** *A trace-high clip.*

resistance of the animal, and of achieving favourable results, are the head, groin and belly regions; these are best left to the last. The blades of the clipper should lie parallel to the coat and should not dig into the horse. They should be used against the lay of the coat.

On no account must hair be removed from the inside of the ears as it provides natural protection.

When dealing with the legs it is as well to use a clipper head with a coarser lower blade, called a 'leg plate', as it does not cut off the hair so close to the skin.

Care must be taken to clip neither the sides of the mane nor the root of the tail. The practice of removing a short portion of the mane in the wither region is not recommended. If a portion of the mane is removed to make way for the head-piece of the bridle, it should be only slightly broader than the head-piece.

As clipping proceeds, throw a folded rug over the loins: a cold horse soon gets fidgety. While clipping parts other than the head allow the horse to feed from a haynet so that he will not get bored. Unless the whole leg is being clipped the use of clippers on the back of the tendons or fetlocks is not recommended as the appearance of the legs will be spoiled for some weeks. When the hair does not pull easily, trimming-scissors and a comb may be used instead. The comb and scissors must be moved upwards, i.e. against the hair in the same way that a hairdresser trims a head of hair.

If a horse is clipped out, i.e. his natural coat removed, it must be replaced by from one to three blankets and a rug, depending on the weather. To leave him without a rug is unkind, and he will not thrive.

When to Clip

Some time during September it will be noticed that for the first time for many months the coat appears dull and rough. This indicates the onset of the change of coat, and from then on the winter coat begins to grow: hence the saying 'no horse looks well at blackberry time'.

Sufficient time should be allowed for the new coat to become well established before its removal is contemplated. Horses vary greatly as to the time at which they require their first clip, and some grow a much thicker coat than others.

The first clip of the season is usually made in October. Thereafter, the coat continues to grow, although never as fully as before, so that more than one clip may be necessary. The second or third clip, however, should not be delayed later than the last week of January.

TRIMMING

Trimming is the process of tidying up a horse and includes among other things the pulling of the mane and tail.

The hairs of the mane and tail will pull more easily when the pores of the skin are warm and open, i.e. after exercise or on a warm day. Many horses will fidget, and if pulling is done in very cold weather when the pores of the skin are tight shut it may cause pain. Pull out only a few hairs daily to avoid soreness.

The Mane

Pulling To thin out an over-thick mane, to reduce a long mane to the required length, or to allow the mane to lie flat. The longest hairs from underneath should be dealt with first and removed a few at a time. Do this with your fingers or by winding a few hairs at a time round the comb and plucking them out briskly. Never pull the top hairs, nor any hairs that may stand up after plaiting, because they will form an upright fringe on the crest. On no account must scissors or clippers be employed.

Hogging This involves complete removal of the mane by means of clippers. It is done when a horse grows a ragged mane that spoils its appearance, or to avoid the work involved in the care of a mane. Hogging can only be done neatly by getting an assistant to stand in front of the horse and, taking hold of his ears, gently forcing his head down so that the crest is stretched. Clippers are then used to remove the whole of the mane, beginning at the wither end and working upwards towards the poll. Care must be taken that no unsightly line is left where the coat meets the crest. Hogging needs to be repeated about every three weeks.

Once a mane has been hogged, re-growth will take at least two years to look neat and even then the mane may not regain its former appearance.

Plaiting is done for neatness, to show off the neck and crest, and to train the mane to fall to the side preferred, normally the off side of the neck. There should always be an uneven number of plaits down the neck, plus the forelock. The minimum number of plaits for a hunter including the forelock is six. There are several ways of plaiting a mane.

The following are two of the easiest methods:

First method. You will need: a water brush, some pieces of thread about eight inches long, a needle with a large eye, a mane comb and a pair of scissors. First damp down the

53 *Mane plaiting, showing use of needle and thread.*

mane with a wet brush and divide it into five or more parts. The plaiting of each division should now begin. When you are three parts of the way down, take a piece of thread, double it over and plait it in. When a plait is complete, the ends of the thread should be looped round the plait and pulled tight. Having completed all the plaits, ensure that they are of nearly even length. If one is considerably longer than the others make a further loop over. Now take the needle, pass both ends of the thread through the eye, and doubling the end of the plait under, push the needle through the plait from underneath and close to the crest. Pull the ends of the thread through, remove the needle and bind the thread tightly round the plait. Finish by knotting the thread on the underside of the plait and cutting off the protruding ends with scissors *(figure 53)*.

Second method. A rubber band is used instead of a needle and thread. Having finished a long plait, loop the band several times round the end. Then turn under the end of the plait into the required position and loop the band around the whole until tight.

Mane plaits should not be left in all night as they tend to pull and be uncomfortable.

The Tail

A long tail collects mud, is likely to become straggly, and hides the hocks. A show horse with a high head carriage needs a longer tail than a hunter.

Clippers should never be used on the tail except when 'banging', nor should a dandy brush be used. A mane comb breaks and tears the hair and should therefore only be used to help you when pulling a tail.

Pulling A well-pulled and tidy tail adds greatly to appearance. Tail pulling *(figure 54)*, should not be carried out on a horse living at grass, as this deprives him of natural protection in the dock region. You will need: a body brush, a water brush, a mane comb, trimming scissors and a tail bandage.

First groom the tail well to remove all tangles and to separate out the hair. Begin pulling at the dock region by removing all the hair from underneath. Then work sideways, removing the hair evenly on both sides of the tail. Remove only a few hairs at a time—either with your fingers or by winding the hairs around a comb and giving them a sharp pull. A little resin on the fingers may help. Further procedure depends upon the type of tail desired.

A bang tail. The end is cut off square 10–20cm (4–8in) below the points of the hocks. Cutting is done with the help of an assistant, who places his arm beneath the root of the tail: the cut is then square when the tail is carried naturally.

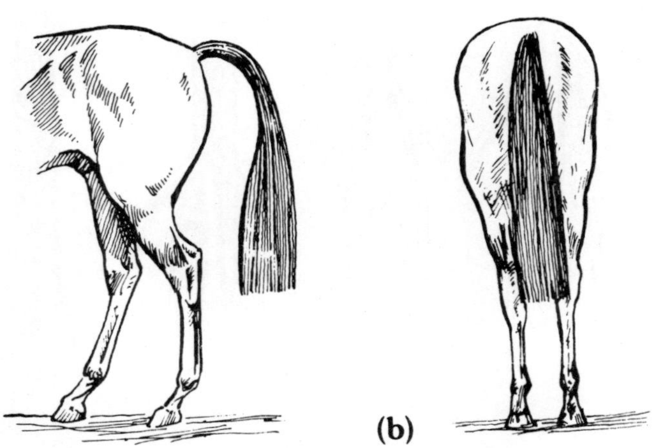

(a) (b)

54(a) *A pulled tail.* **(b)** *A full tail (unpulled).*

A switch tail. Here, pulling is continued drastically to the extent of about half the length of the tail. The ends of the tail are allowed to grow to a natural point.

After pulling, the tail should be bandaged in the manner described on page 122. The regular use of a stockinette bandage on the dock region greatly helps the preservation of the shape.

Plaiting This is a useful alternative to tail pulling for horses who resent that procedure, or who have untidy, bushy tails, and who are kept at grass. Plaiting the long hairs of the dock region makes for neatness and tidiness without depriving the horse of the protection that a full tail affords. The procedure is as follows:

The tail hairs at the dock are allowed to grow long. A small number are separated with the finger and thumb on either side of the tail and knotted together with thread. This knot hangs down the centre of the tail, and successive small bunches (8 to 10 hairs) from either side are plaited with it. Alternatively, a small number of hairs are taken from the middle of the dock *(figure 55)* and again successive bunches from either side are plaited in with them. Plaiting in this manner continues downwards for about two-thirds of the length of the dock. Thereafter, plaiting is confined to the centre hairs of the tail, and the knot is incorporated so that a free-hanging 'pigtail' is formed. When the end of the pigtail is reached, it is secured with thread

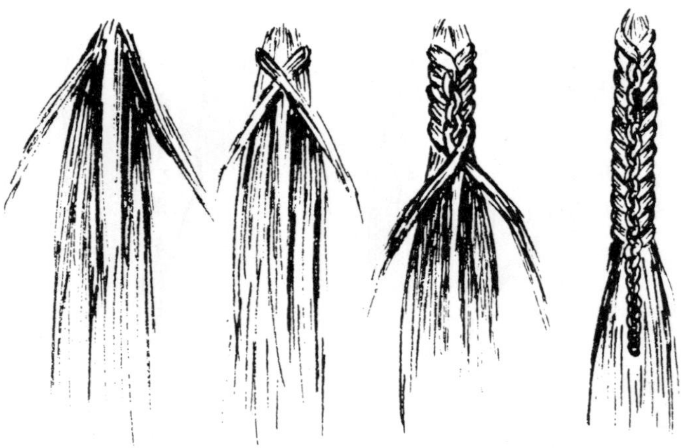

55 *One method of plaiting a tail.*

and looped back under itself to the point where the side hairs ceased to be included. The resulting loop is stitched together to form a double-thickness plait.

The Coat

In the early part of the year when the winter coat has stopped growing on a clipped horse, long hairs—known as 'cat hairs'—start showing, giving an untidy appearance. These are best removed with a clipping machine. Long hairs which grow in the jowl region and at the back of the tendons may be removed, a few at each grooming, by plucking with the fingers.

Feeding

Feeding is an art. Grazing is the natural method, and grass the natural food. When a horse is stabled and in work, it is necessary to substitute for this natural food a dry and concentrated diet. The whole art of feeding the stabled horse lies in the successful adjustment of the natural to the artificial diet. Our knowledge of how to make this adjustment has been acquired over many generations as the result of practical experience and scientific study.

You should not be content merely to know the right way in which to feed, but should also have some understanding of why this way is the most satisfactory. A horse has a remarkably small stomach, nature's intention being that he should have a small amount of food in him all the time, but never a lot at any one time. As a result the natural method is to eat a little at a time throughout most of the day and night. When a horse is brought up from grass, hay replaces grass as bulk. Concentrated foods must be introduced only gradually.

It is possible to deal here only with the correct system of feeding for the majority of horses. Others, such as 'bad doers', shy feeders, or gross eaters, will require special treatment.

RULES OF GOOD FEEDING

The rules which every good horsemaster should follow, are:

- Feed little and often. Imitate the natural method as far as possible.

- Feed plenty of roughage (hay). This is necessary so that, as in grazing, the digestive organs are always well filled. A successful digestive process in the horse is impossible without adequate roughage.
- Feed according to the work done and to the size and temperament of the horse. Increase the amount of concentrated food (oats, etc.) if the demands of work are heavy; reduce the amount if they become light; stop giving concentrated food if the horse has to be laid up, but remember to increase the bulk food in order to compensate for the lack of concentrates.
- Make no sudden change in the type of food or in the routine of feeding. All adjustments must be gradual and spread over several days.
- Keep to the same feeding hours each day.
- Feed clean and good quality forage only. The horse is a fastidious feeder and will relish only the best. Musty and dusty food not only adversely affect condition but often actually prove harmful.
- Feed something succulent every day if possible. For example, give the horse green food or carrots to compensate for the lack of grass.
- Do not work a horse fast immediately after a full feed, or when the stomach is full of grass. The stomach lies next to the chest and will press on the lungs when it is full thus affecting the horse's breathing.
- If water is not constantly available in the stable, water before feeding. Undigested food will not then be washed out of the stomach. When water is kept continuously in the stable, a horse will frequently take a short drink during, or after, a feed. This will do him no harm so long as it is not a lengthy draught.

FORAGE

Concentrated Food

Oats Experience has shown that oats are the best all-round food for horses, but they must be fed sparingly to ponies who respond rapidly to concentrated food, may become excitable ('hot up'), and therefore be difficult to manage and to ride.

The grains should be large, hard and clean. Oats may be fed

whole but are easier to digest if bruised, rolled or crushed. Once having been treated in this way, oats should not be stored for more than about three weeks since they will go stale and lose their nutritional value. This will also happen if they are rolled or crushed too severely. Oats are best fed with chaff (of hay or straw), or a little bran, or with a mixture of both.

Cubes These are a carefully formulated mixture of many ingredients which include added vitamins. There are several varieties and the nutritional value of each is consistent—they vary in this value only according to their type: 'Horse and Pony', 'Racehorse', 'Stud', 'Transit', 'Grass', etc. The well-known proprietary brands are reliable, but some of the locally-made ones may not be of the same quality. Cubes may be used as a substitute for part or all of the oat ration—about 340–680g ($\frac{3}{4}$–1$\frac{1}{2}$lb), replacing approximately 450g (1lb) of oats depending on the type of cube and the manufacturer's recommendations. (Cubes are sometimes referred to as 'nuts'.)

The feeding of cubes has many advantages; it saves storing several different kinds of grain; it saves mixing feeds; and it ensures that the horse gets a standardized, balanced diet, with the necessary vitamins and minerals. Also, especially with 'Horse and Pony' cubes, the animals are less likely to 'hot up' than when only oats are fed. Disadvantages are that cubes are quite expensive, rather boring as a diet and tend to deteriorate when stored. When the feed consists only of cubes, adjustment of the diet is not possible as it is with traditional feeding. Care must also be taken to feed the correct type of cube, as there could be danger in feeding too much high protein to certain types of horse and pony. The addition of chaff or bran will ensure adequate mastication and salivation to moisten the cubes before swallowing.

Barley Rolled, crushed, or preferably flaked, barley can be used as a substitute for oats. It has similar food value to oats but is not so likely to cause a horse to 'hot up'.

Boiled barley Fed warm and mixed with bran, it is a useful food, especially after hard exercise such as hunting. It will often tempt a shy feeder to eat and can be used to add variety to the diet of a stale or over-worked horse, and for fattening purposes. It must be brought to the boil and simmered until the grains split (four to six hours).

Flaked maize This is fattening and should be used sparingly as it may cause over-heating of the blood.

Wheat Except as a bran, wheat is not suitable for horses.

Bran Wheat bran may be used to encourage mastication and digestion, and to provide bulk; it is invaluable in the feeding of invalid horses. The milling process is so efficient nowadays that a good sample of bran, i.e. one with some flour remaining in it, is difficult to come by. It is also very expensive. Fed in excess it can seriously affect both growth and health.

Bran mash is a very useful warm food after hard exercise and hunting. To make a bran mash, fill a bucket three-quarters full of bran; pour boiling water over it until it becomes thoroughly wet; stir in 28–113g (1–4oz) salt and a handful of oats or boiled barley; then cover with a sack and leave until it has become cool enough to eat. A little linseed jelly mixed into it adds to the taste and is beneficial to the horse. Bran mash has a laxative value, and it has everything to recommend it when fed once a week to horses in work and to invalid horses. It is also a convenient way of administering medicines such as worming compounds.

Beans When dried, beans are very nutritious but very heating—too much so for ponies except when living out in winter. At this time if fed sparingly they may be beneficial. A double handful twice a day mixed with the other feed is sufficient. Beans should be bruised or split before feeding.

Linseed The high oil content of linseed improves condition and gives a gloss to the coat. It is generally fed to horses only during the winter months. It can be prepared and fed either as jelly or gruel-tea. The correct allowance is a twice weekly feed of 113–226g ($\frac{1}{4}$–$\frac{1}{2}$lb) of the seed (weighed before cooking). Horses cannot absorb more than this amount daily.

Linseed jelly A special saucepan should be set aside for preparing this. Place a handful of linseed in the saucepan, cover it with water and allow it to soak with a lid on until the next day. A cool or only slightly warm oven is the best place for this soaking process. The next evening add more water and bring to the boil. It is important that the water should boil, as unboiled linseed is poisonous. Set the saucepan aside and allow the contents to cool. If properly made, the linseed should set as a jelly which is then mixed with the evening feed.

Linseed gruel (or tea) The preparation of linseed gruel (or tea) is the same as for linseed jelly except that more water is used. The water in which the linseed is cooked is highly nutritious, and is used with bran to make a linseed mash.

Oatmeal gruel This is useful as a pick-me-up for a tired horse on return from work, though not all horses will take it. Place a double handful of oatmeal in a bucket, pour on boiling water and stir well. Leave it to cool before offering it to the horse. Gruel prepared in this way must be thin enough for the horse to drink.

Molasses A by-product of the manufacture of sugar, it is of high nutritional value. Being tasty it is useful in persuading shy feeders to eat other foods, but on no account must it be used to persuade horses to eat inferior food. In its liquid form it resembles black treacle. Dilute up to two tablespoonsful and sprinkle them over, or mix them in with, the feed. Alternatively, buy it in meal form—'Molassine' is one type; in this case a handful per feed is sufficient.

Dried sugar beet pulp As a source of energy and roughage, dried sugar beet pulp can usefully be included in a concentrated diet to maintain weight and to provide heat and bulk for a horse not doing fast work. If fed in excess it has high laxative qualities.

Dried sugar beet pulp must always be soaked in cold water for 12 hours before being fed to a horse, because in its dry state it will cause choking or may swell in the stomach and cause colic. Sugar beet pulp should be prepared in a ratio: $2\frac{1}{2}$ parts cold water to 1 part pulp. No more than 1.4kg (3lb) (soaked weight) should be fed per day. Once soaked it should be fed immediately, as the fermentation process which begins soon after it is mixed with water will be harmful. Sugar beet should not be soaked in hot water, as this increases the speed of fermentation.

The pulp can be purchased in cubes which, being densely compressed, should be soaked for longer and in a greater quantity of water. Sugar beet pulp cubes should be prepared in a ratio: 4 parts cold water to 1 part cubes. Both dried pulp and cubes should be stored in a cool, dry place, as mould quickly forms on them. Take care to differentiate between cubes of sugar beet and the other types of cube.

Salt This is an essential part of a horse's diet. It is best provided either as a salt lick in a special container fixed to the stable wall, or as a lump of rock-salt in the manger; in either case the horse is able to have a lick whenever he feels he requires salt. Alternatively, 28–113grm (1–4oz) of table-salt can be added to the feed once a day.

SUCCULENT FOODS

The good feeder should be constantly on the look-out for something succulent, i.e. green or juicy, to add to the horse's diet. This makes the feed more appetising, provides bulk and variety, and in part satisfies the natural craving for grass, as well as providing valuable vitamins.

Green foods such as grass, lucerne, etc., should be easily obtainable in summer and every effort should be made to provide them. They may be fed either in a haynet or 'chaffed up' with hay and added to the feed.

Carrots Particularly acceptable during winter months, carrots are highly nutritious. Some horses also relish swedes, mangels, turnips, beetroots, or parsnips. To prepare these roots, first scrub well under a running tap, the warm water tap if preferred. Then slice the roots lengthwise into long 'fingers' and mix with the feed. Square or round pieces of root must not be fed as they are liable to become lodged in the throat and so choke the horse. Begin by feeding 0.45kg (1lb) a day, and increase the amount to 0.9kg (2lb), or even more, when the horse is accustomed to them.

Apples These are always especially appreciated.

ADDITIVES AND FOOD SUPPLEMENTS

There is a fairly wide variety of proprietary additives and food supplements on the market, of varying content and nutritional value. Consult a vet before using them and on no account mix different brands without taking his advice, since the ingredients may react against each other and cause serious disorders.

BULK FOODS (ROUGHAGE)

Hay

This provides all the bulk needed as a substitute for grass. Plenty of good quality hay should always be available for the stabled horse and for a horse living out during the winter.

Feeding bad hay is a false economy as the horse will not thrive on it. It is best not to feed hay that is less than six months old, as this may cause digestive upsets. Hay containing weeds such as docks and thistles is best avoided, and any containing ragwort should be discarded. Mouldy hay should never be fed—it may cause irreparable damage to the horse's respiratory system.

It is better to feed good quality meadow hay than bad or indifferent seed hay.

Types of Hay

Seed Hay This comes from re-seeded land and is also known as 'mixture' hay. It is an excellent food for horses, and contains good grasses such as rye, clover, timothy, sanfoin, cocksfoot, meadow fescue, etc. A good sample of seed hay should be greenish brown, hard and crisp to the feel and sweet to smell. A yellow or dark brown colour usually denotes deterioration. A musty smell indicates hay that has been harvested when damp; this is not good for horses.

Meadow Hay This comes from pasture permanently laid down to grass; the stalks are less coarse than in seed hay, and are thus not so good for mastication. Its nutritional value may vary considerably, depending on the types of grass it contains. Meadow hay should be greener than seed hay, soft and sweet smelling.

Chaff or Chop Hay, either by itself or mixed with a small proportion of oat straw and any green food available is passed through a chaff cutter and then fed with the corn feed. It adds bulk to the oat feed, ensures better mastication, and prevents the horse from bolting the food. Chaff cutters form part of the equipment of all high-class stables.

'Horsage' or 'Haylage' This is a cross between hay and silage, and is particularly useful for feeding to horses who are allergic to hay or straw. It is fed in similar quantities to hay. Care should be taken when feeding the high-protein variety.

Straw and Other Bulk Foods

Oat straw This sometimes makes up a proportion of the bulk food used either on its own or in conjunction with hay when making chaff. Barley and wheat straws are not suitable as feed.

Straw may provide up to one-third of the roughage fed and is better than mouldy or very poor hay, but its nutritional value is small.

Silage

This is occasionally fed to horses, but must be introduced gradually. It may provide up to about one-third of the hay ration. Care should be taken to avoid the high-protein silage that is often fed to dairy cattle.

FEEDING ARRANGEMENTS

Feeding arrangements should be planned in accordance with the rules set out on pages 143 to 144.

First you must decide how much and what type of feed the horse needs. This depends chiefly on whether he is stabled or at grass, on the amount of work he is doing, on the varieties of forage available and on his size. In addition, you must make provision for the horse to receive sufficient bulk food each day.

Having settled these points, work out a routine that takes them, and your other commitments, into account. The programme must be arranged so that the horse is fed at least three, and possibly four, times a day, always at the same hour, or that he has access to grazing and can feed himself.

In addition, it is important to bear in mind the following three points:

- It takes a horse twenty minutes to eat a full corn feed, and one and a half hours to digest it. If, therefore, a horse is to be fed before work, it must be done one and a half hours beforehand or he must be given only a small feed.
- It takes a horse two hours to eat 3.6kg (8lb) of hay, so the bulk of the hay ration is best given after work and at night when the horse has time to eat and digest it quietly.
- Horses thrive on routine and in particular on regular feed times, so it is better not to plan feed times that you cannot stick to regularly. For example, do not include late evening feeds if you are likely to be out frequently at night. If after a reasonable time the horse has not finished his feed, remove what he has not eaten. Never leave stale food in the stable.

Making Up and Giving Feeds

Feeds should be made up and mixed in a container before being taken to the horse. A scoop of known capacity should be used so that quantities of oats and other foods can be accurately gauged.

When feeding bran, dampen it slightly before mixing it with the other foods.

A stabled horse Most conveniently fed from a manger, but if one is not available, use a metal or heavy rubber feed tin on the floor. This should be removed from the box after the horse has finished his feed. Whichever is used, it must be kept clean and should be washed out each day. Wasteful feeders, that is to say horses who throw their feed out of the manger, may be defeated in one of the following ways: by keeping a brick, large round stone or, best of all, a lump of rock salt in the manger; by having bars fitted across the manger; or by using a nosebag.

A horse at grass He should be fed either from a bowl or feed tin of metal or heavy rubber, from proper feed boxes on the ground, or from a manger hung over the fence or gate *(figure 56)*. There should be one feed tin or box for each horse in the field. Tins and boxes should be well apart, and kept clean. New pastures tend to lack the deep-rooted herbs which are an essential part of the diet; to counteract this, extra food or food supplements must be provided.

56 *Feed tin, feed box and portable manger.*

Feeding Hay

The natural feeding position is from ground level, but this is wasteful since much of the hay will get trampled and soiled. Hay is best fed to the stabled horse from either a hay rack or a haynet fitted or hung at about eye level. If higher than this, the horse is likely to get seeds and dust in his eyes, and if lower he may catch his feet in it when rolling.

When a haynet is used for feeding a grass-kept horse it should be attached well clear of the ground to either a fence, a post or a tree, and tied as illustrated *(figure 57)*.

A haynet is probably the most economical method of feeding hay, both in the stable and in the field. It also has the advantage of being easily weighed.

57 *Two ways of tying a haynet.*

Storing Forage

If possible, store forage away from the stable. Concentrated foods in particular must be safeguarded; horses have died of colic through over-eating, after having got loose and into the food supply. These foods should be kept in bins which are proof against vermin and not in sacks. The bins should have either secure catches or lids that are heavy enough to prevent the horse raising them.

Hay and straw should be stacked on wooden slats that allow the air to circulate underneath and prevent damp from rising into the bottom bales.

FEED SCALES

The following suggestions are intended as a guide only, and should be read in conjunction with the general notes on feeding. They should be interpreted with common sense and imagination. Feed scales must be adjusted to take into account a variety of circumstances such as:

- The particular type, temperament and age of the horse.
- The condition and fitness of the horse.
- The work being done.
- The weather.
- The time of year.
- A naturally good feeder or a bad 'doer'.
- A clipped, partially clipped or unclipped horse.
- The type and number of rugs—including New Zealand rugs when at grass.
- The grass available, etc.

As a rough guide, the average 16-hand horse requires a total weight of from 11.3–13.6kg (25–30lb) of food per day. When in strong work, this might be made up of 4.5–6.3kg (10–14lb) of concentrates, and the balance in hay and other bulk foods.

Some people prefer to relate the total daily weight of food to the horse's weight on the basis of approximately 1.2kg ($2\frac{1}{2}$lb) of food per 45.4kg (100lb) of body weight per day. There are various formulas for estimating a horse's weight in relation to his girth and length other than by weighing a horse on a weigh-bridge or farm weigh-crate. Variations in weight can be a useful monitor of growth, health and training programme, and a comparison can be simply made by measuring his girth circumference behind the elbows and noting variations week by week.

Remember that much of a horse's food goes to keeping him warm, so while fresh air is essential, the more you can reduce loss of body heat by providing a draught-free box, or a shelter outside, and by rugging up sensibly and efficiently, the less food he will require.

Horse in Full Work:

Stabled, clipped, rugged	*16.2hh*	*15.2hh*	*14.2hh*
Oats or Barley (and/or Cubes—see Notes)	5.4–6.3kg (12–14lb)	4.5–5.4kg (10–12lb)	1.8–3.6kg (4–8lb)
Chaff and/or Bran and/or Dried Sugar Beet pulp (After soaking)	0.9–1.8kg (2–4lb) 1.3kg (3lb) only	0.9–1.8kg (2–4lb) 1.3kg (3lb) only	0.9–1.3kg (2–3lb) 1.3kg (3lb) only
Hay	6.3kg (14lb)	5.4–6.3kg (12–14lb)	4.5–5.4kg (10–12lb)

Notes

Cubes May be substituted for oats in part or as a whole at the rate of 340–680g (¾–1½lb) cubes per 452g (1lb) of oats, depending on the type of cube and the manufacturer's recommended rate.

Bran Expensive and frequently of poor quality. A double handful per feed is normally sufficient. Its function of providing extra bulk and roughage and breaking up the corn feed is equally well done by chaff or cubes, or dried sugar beet pulp. (Feed no more than 1.3kg/3lb of the latter, soaked for 12 hours.)

Bran/Linseed mash After hard work, or once a week.

Beans These add variety, but are not essential. 453g (1lb) per day in place of a similar amount of oats are adequate.

Flaked Maize As for beans (above).

Molasses 2 tablespoonsful watered down and sprinkled on the feed, or

Molassine meal One handful per feed. Neither is essential, but both are of high nutritional value and are tasty.

Carrots, apples, etc. Feed up to 0.9kg (2lb) per day.

Salt As a salt lick. Alternatively 28 – 113g (1 – 4oz) can be added to the feed once a day.

Additives and supplements As advised by the vet.

Horses turned out to grass They should receive 0.9–1.8kg (2–4lb) of oats or equivalent in late summer. On being 'brought up', they should start on this amount, and the amount should be built up gradually as their work increases.

Horses working from grass or 'out-by-day, in-by-night'
They should receive the same or, if the weather is severe, slightly more concentrates than the stabled horse, the quantity of hay being adjusted to suit the amount and quality of the grass available.

Child's Pony:
Taking part in Rallies, Shows, Hunting, Gymkhanas, etc.

	Pony 12.2 to 13.2hh		Shetland Pony
	Summer At grass day and night	*Winter* Stabled at night, at grass by day	*Winter* At grass day and night
Cubes	1.8kg (4lb)	1.8kg (4lb)	226–452g (½–1lb)
Bran	226–452g (½–1lb)	226–452g (½–1lb)	226–452g (½–1lb)
Hay	—	2.3kg (5lb)	2.7kg (6lb) Plus carrots

- The above figures, intended purely as a guide, are very approximate, bearing in mind that it is better to under-, than over-feed concentrates to ponies. A Shetland pony in summer requires no feed other than grass.
- Bran is not essential and, unless of good quality, should be omitted.

Watering

The importance of a clean and constant supply of water cannot be too strongly emphasised. Nothing affects condition so quickly as faulty watering arrangements.

IN THE STABLE
The most satisfactory way of providing water is to use either buckets or an automatic drinking bowl.

Buckets

These are probably the most practical. They are best made from rubber, plastic or polythene which are lightweight, noiseless, and will not cause injury to the horse. Two buckets may be necessary to ensure a constant supply of water. They should be placed, preferably but not necessarily, in the corner of the box, either on the floor or suspended from a special ring fitting (see page 101), away from the manger and the hay where they may become soiled, and away from the door where they may get knocked over. If possible they should be within sight of the door so that their contents can be easily checked.

Automatic Drinking Bowls

These are filled either by gravity or by the horse himself pressing on a lever with his nose. Not all horses will use them; and their shallowness which may prevent a deep, full drink is a drawback. Special plumbing is necessary and each bowl must have its own stop-cock. Unless inspected daily they are liable to become clogged.

AT GRASS

Rivers and Streams

When a horse is kept at grass it is a great advantage if the field has a stream passing through it. Ideal conditions are provided by a river or stream with water running over a gravel bottom and with a good approach. Drinking from a shallow stream with a sandy bottom may result in the horse taking up a small quantity of sand each time he drinks. If the sand accumulates inside him it may cause an attack of sand colic. Streams with steep banks or a deep, muddy approach are unsuitable watering places.

Ponds

If the only water supply available in a field is a pond of stagnant water, alternative arrangements should be made.

Field Troughs *(figure 58)*

Field troughs filled from a supply of piped water provide the best arrangement for watering the horse kept at grass. Galvanised iron troughs are excellent. They should be approximately

A self-filling trough with the ball-cock apparatus in an enclosed compartment.

Another good arrangement. The tap is positioned so that the horse cannot interfere with it; there is no projecting inlet pipe on which he can become caught up.

A bad arrangement. The sharp edge of the bath may lead to injury to the knee, and the projecting tap is dangerous.

58 *Field troughs connected to a supply of piped water.*

1–2m (3–6ft) long, about 38cm (15in) deep and placed so that the top is about 60cm (2ft) from the ground. They must be provided with an outlet at the base for emptying. Troughs should be placed clear of trees so that leaves do not accumulate in them. The ground should be well-drained, otherwise the earth surrounding the trough will become muddy in winter. The nearer the trough is to the house, the gate or the road, the better the chance that it will be regularly inspected. During periods of frost and snow, troughs should be attended to at least twice a day.

Troughs are best filled by means of a ball-cock apparatus enclosed in a covered compartment at one end. The trough then automatically fills whenever the horse takes a drink. If a tap is used it should be placed at ground level and the piping fitted in such a way that it hugs the side of the trough and has no projections on which the horse may injure himself. A projecting tap or stand-pipe in the vicinity of the trough is extremely dangerous. There should be an outlet hole in the side of the trough positioned so that the water level remains below that of the inlet pipe.

Field troughs require regular inspection, emptying and cleaning. After this has been carried out it is sensible to leave them empty for a few hours to allow the air to sweeten them.

WHEN TO WATER

As has already been stressed, horses should, if possible, have a constant supply of clean, fresh water available to them both in the stable and at grass. Knowing water is available, a horse will drink when necessary and take no harm.

If a horse has been allowed a full drink while at work, he should not be put in a fast pace for at least half an hour. He will suffer no ill-effects if he drinks from a trough with the ice broken up.

A horse that is hot and sweating will seldom come to harm if he takes a full drink of cold water, provided that he is walked about afterwards.

On return from strong work and after a long time without water, it is advisable to give the horse a bucket half full of water with the chill taken off—21°C (70°F)—to quench his immediate thirst, and repeat this a quarter of an hour or so later. He will then be less liable to break out into a sweat.

In the course of a long day's hunting a horse may with advantage be allowed a short drink at a convenient watering place, but a full draught should not be allowed if fast work is to follow immediately afterwards. The same applies when hacking home after hunting.

The Foot and Shoeing

Every horseman should have some understanding of the care of a horse's feet and of shoeing. 'No foot—no horse', is an old and very true saying.

THE STRUCTURE OF THE FOOT *(figure 59)*.

The Exterior of the Foot

This consists of three parts: the wall, the sole and the frog. All three are horny structures and are non-sensitive, with neither nerve nor blood supply. This explains why shoeing nails can be driven through the wall and why the frog and sole can be cut with a knife without causing pain or bleeding.

The wall This is the part of the hoof visible when the foot is on the ground. It grows downwards from the coronet, just like a finger-nail, and in the natural, unshod state the rate of growth equals the rate of wear. The wall encircles the foot and at the heels is inclined inwards to form the bars.

The outer surface of the wall has a glossy, varnish-like finish which prevents undue evaporation from the horn so that it does not degenerate and become hard and brittle.

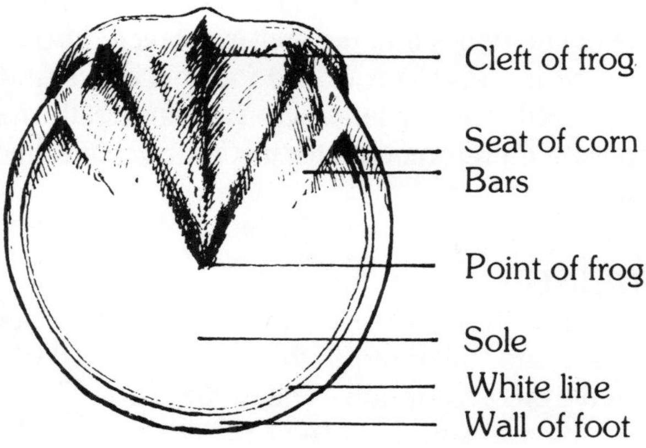

59 *The parts of the foot.*

The toe, the quarters and the heel all form part of the wall of the foot.

The sole This protects the foot from injury from below. As it is none too thick for this purpose, liberties cannot be taken with it. In a healthy state it is slightly concave, i.e. like a saucer turned upside down, and thus helps to give a better foothold.

The frog Nature's anti-slipping and anti-concussion device—it is the first part of the foot to make contact with the ground and thus is important in ensuring a good foothold. Its peculiar wedge shape, irregular surface, and central cleft help its anti-slipping function. Its effectiveness as a shock-absorber stems from its size and india-rubber-like consistency, its upward flexibility and the cushion within the foot upon which it rests.

The importance of a healthy frog cannot be over-stressed, when it is remembered how much lameness is directly attributable to slipping or to concussion in the leg.

The Interior of the Foot

The inner foot is made up of bones, joints and sensitive structures, any or all of which are liable to injury should the wall, sole or frog be penetrated.

SHOEING

The need for shoeing is the direct consequence of domestication, i.e. the working of a horse on hard roads which causes the wall to wear away at a greater rate than that at which it grows. The effect of shoeing is that the wall is protected from all wear, but since it continues to grow, the foot becomes unduly long and, if neglected, chronic lameness or even death may result. The shod foot, therefore, calls for just as much attention as the unshod foot.

RE-SHOEING

As a general rule, a shod horse should visit or be visited by the farrier once a month. Even if shoes are not badly worn the horn will have grown enough to need reducing. This is particularly true of the toe region, and a horse with an over-long toe is always liable to stumble.

The procedure by which a slightly worn shoe is taken off, the foot reduced in length and the same shoe replaced, is known as a 'remove'.

A horse subjected to heavy work on hard roads may wear his shoes through in less than a month and in this case visits to the farrier need to be more frequent.

The indications that a horse is overdue for re-shoeing are:

- The shoe is loose.
- Some part of the shoe has worn thin.
- The clenches have risen and stand out from the wall.
- The foot is over-long and out of shape.
- A shoe has been 'cast', i.e. lost.

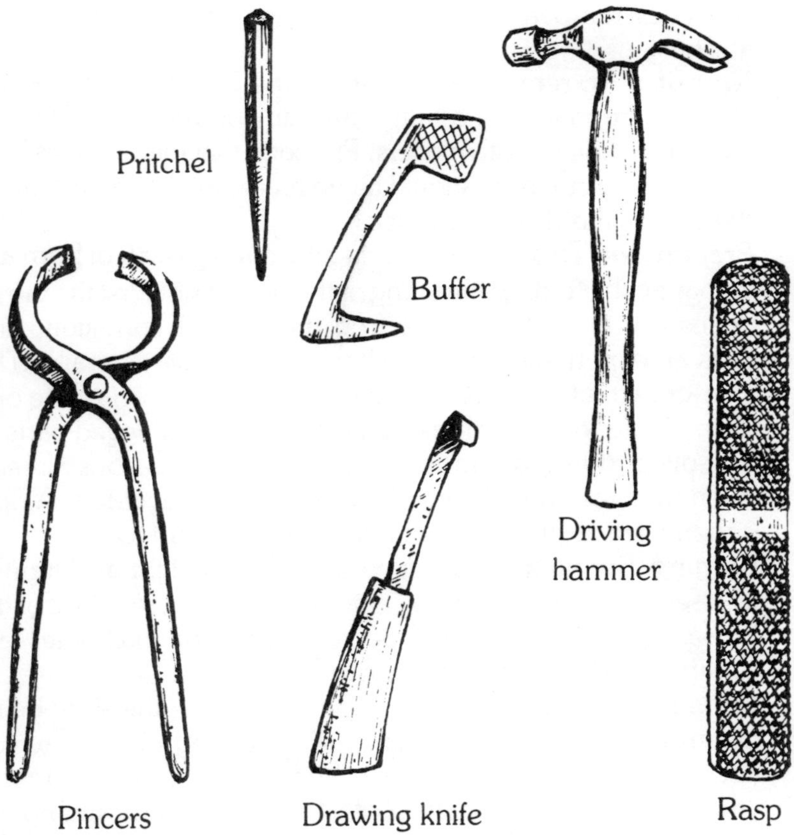

60 *A farrier's tools.*

SYSTEMS OF SHOEING

There are two systems, namely 'hot' shoeing and 'cold' shoeing. In 'hot' shoeing the shoe is specially made to fit the foot. It is tried on hot and adjustments are made before it is finally nailed on. For this method a visit to the farrier may be necessary, though many farriers now have portable forges. In 'cold' shoeing, a ready-made shoe is fitted and adjusted as much as is possible on a cold shoe. In this case it is not necessary to take the horse to the forge. It will be easier if you can make arrangements for the farrier to come to you. The former system is the more satisfactory one.

Hot Shoeing

The procedure for re-shoeing a horse by this method falls into six stages: removal; preparation; forging; fitting; nailing on; and finishing.

Removal To remove an old shoe, the farrier first cuts all the clenches by means of a buffer and driving hammer. He then levers the shoe off with pincers. Provided that the clenches have been cleanly cut there should be no breaking or tearing away of the wall as the shoe is released.

Preparation This entails reducing the over-growth of horn and preparing the foot for the fitting of the new shoe. First the farrier cleans out the sole and the frog, notes their condition, and casts an expert eye over the shape of the foot generally. The over-growth of wall is removed with either a drawing knife or a toeing-knife, or sometimes with a horn-cutter. Ragged parts of the sole and frog are trimmed away, but for reasons already given, unnecessary cutting of them should be avoided. A rasp is then used to give the foot a level bearing-surface.

Forging The making of a new shoe. The weight and type of iron selected depends on the nature of the work required of the horse. After the iron has been shaped the nail-holes are stamped and the clips are 'drawn'.

Fitting Carried out while the shoe is still hot—the shoe being taken to the foot on a pritchel. The searing of horn which occurs when the shoe is fitted on the foot indicates the extent to which foot and shoe are in contact, but searing should not be overdone. After searing, any adjustments necessary to the shape of the shoe or the length of the heel are made.

Nailing on The shoe is cooled by immersion in water and nailed on—the first nail to be driven usually being into the toe. Nails used for this purpose are of special design and have a particular type of head which, correctly fitted, fills the nail-hole however much the shoe may wear away. Nails are made in various sizes and it is important that the right size is used. If too large, the head will project and wear away too soon. If too small, the head will not properly fill the nail-hole. One fault is as bad as the other; both result in early loosening of the shoe.

The end of the nail, where it penetrates the wall, should be turned over and twisted off leaving a small piece called a clench. 'Fine nailing' is when the nail does not penetrate far enough into the wall to give a secure hold. 'Coarse nailing' is when the nail comes out too high up the wall. The risk in the latter case is that the nail may be driven too close to ('nail binding'), or actually into ('pricked foot') sensitive structures.

Clips—generally one toe clip for a fore shoe and two quarter clips for a hind shoe—help to keep the shoe in position and also ensure greater security.

Finishing The clenches are tidied up with the rasp and a small 'bed' made for them in the wall beneath, after which they are 'embedded'. The toe clip is tapped lightly back into position. Finally, to reduce the risk of cracking, the rasp is run around the outer edge of the wall where horn and shoe meet.

The Newly-Shod Foot

When a foot is newly shod, the points to check are:

- The shoe has been made to fit the foot and not the foot to fit the shoe, i.e. that the wall has not been rasped away to meet the iron and that the toe has not been 'dumped'. Either will remove the surface coat of the wall and inevitably lead to cracking and breaking away of the rasped portions.
- The type of shoe is suitable to the work required of the horse.
- The weight of iron chosen is in correct relation to his size.
- The foot has been suitably reduced in length at both toe and heel, and equally so both inside and outside.
- There has been no mis-use of the knife on either the sole or the frog.
- The frog is in contact with the ground.

- The right number of nails has been used. Three on the inside and four on the outside is the usual number.
- The right size of nail has been used, that the heads have been driven home, and that they fill the nail-holes.
- The clenches are well formed, in line, and the right distance up the wall.
- No daylight shows between shoe and foot, particularly at the heel region.
- The heels of the shoe are neither too long nor too short.
- The place for the clip has been neatly cut and the clip itself has been well drawn and well fitted.

A farrier is a highly trained craftsman, so any adjustments should be tactfully requested.

TYPES OF SHOE

Plain Stamped

This is the simplest form of shoe. It consists of an unmodified bar of iron, shaped, stamped with nail-holes and provided with a toe clip. This type of shoe is only suitable for a horse doing relatively slow work since it has no provision against slipping or 'interfering', i.e. knocking or brushing.

Hunter Shoe *(figure 61)*

This is a drastically modified form of the above, designed to meet the needs of a horse moving at a fast pace on grass and pulling up short. This type of shoe is made from concave iron to reduce the risk of suction in soft going and to give a more secure grip on the ground. The part that comes into contact with the ground is 'fullered' i.e. provided with a groove, to ensure a better foothold. The heel of the front shoe is 'pencilled' to avoid the risk of it being caught by the hind shoe and pulled off.

The toe of the hind shoe is 'safed off' (see page 315) and set under the foot to minimise the chances of over-reaching. Quarter clips allow for a 'rolled toe', which is when the metal is drawn up and over the toe of the hoof of a horse which drags its hind feet.

The outer part of the heel is sometimes provided with a calkin giving greater control when pulling up short. The inner heel has a wedge for similar reasons, a wedge being less likely

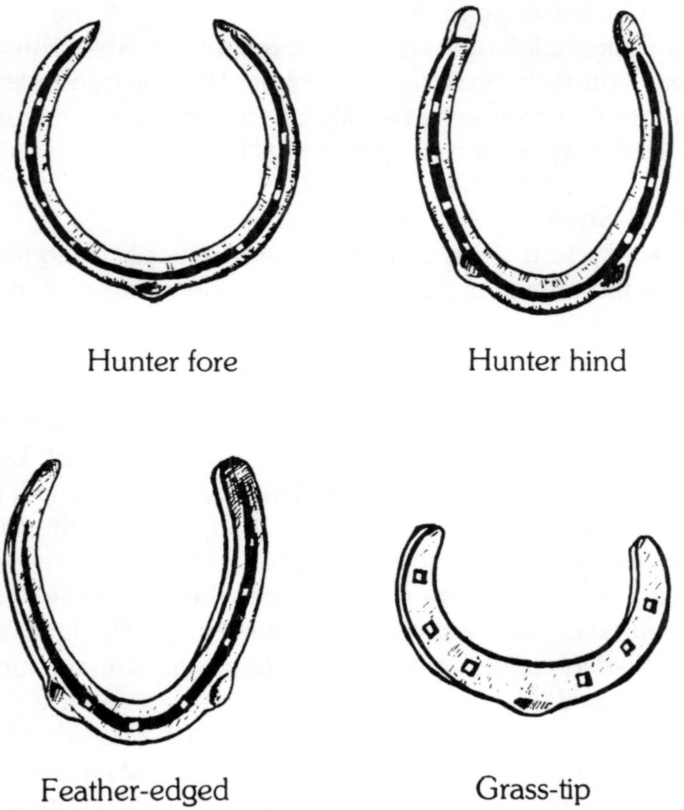

61 *Types of shoe.*

to cause 'brushing' than a calkin. Although primarily intended for the shoeing of hunters, this type of shoe is eminently suited to any horse doing fast work on grass, for example, when showing, jumping and taking part in gymkhanas.

Anti-Brushing Shoe (Feather-Edged) *(figure 61)*
Used for horses that hit and injure the opposite leg, i.e. brushing. The inner branch is 'feathered' and fitted close in under the wall so that the risk of striking the opposite leg is reduced to a minimum. Being slightly higher on the inside, this type of edge has the effect of carrying the foot outwards, which also helps to prevent brushing. There are no nail-holes in the inner branch of this type of shoe.

Grass-Tip Shoe *(figure 61)*
This is a thin, half-length shoe, worn by hunters when running at grass. It protects the wall in the region of the toe from splitting and permits the frog to come fully into action during the run at grass, thus helping to maintain its healthy state.

Surgical Shoe
There are various types of these, each specifically designed to alleviate the effects of injury, malformation or disease in the foot.

Studs
Made of various metals, these can be fitted into the heel of a shoe to lessen the risk of slipping, both on or off the road. To be effective, studs must be of hardened metal except when for use on soft going. Being slower-wearing than the shoe the studs present a rough surface to the ground.

Some studs are fitted permanently into the shoe while others can be fitted or removed easily with a spanner. The latter can be obtained in various sizes and shapes to suit different conditions and needs.

The general purpose road-stud is worn on the outside of the hind shoes. Studs worn on the inside of the shoe increase the danger from 'treads'. By giving the foot too instant a grip, studs worn on the front shoes are apt to put strain on the tendons and should not be used, but in the case of the confirmed 'slipper' on the road, special flat studs or plugs are available. Consult an experienced horseman or farrier before using other types.

Care of studs If riding on the road, do not use soft metal or long studs as the surface will damage them as well as putting strain on the horse's legs. Studs other than road-studs should be removed after use.

When studs are not in use keep them in an oily rag to prevent them from rusting.

Cotton wool inserted in the stud holes when the studs are out helps to prevent them getting clogged up. This is easily removed with a nail. Alternatively, insert a road-stud.

Keep a metal 'tap' with you to clean out the holes in case they do get caked with mud.

WORKING UNSHOD

This is quite feasible provided that work on hard, gritty roads or flinty tracks is avoided. It not only saves on shoeing charges and on visits to the farrier, but an unshod horse has a more secure grip on every type of surface. In addition, injury from a kick by an unshod horse is likely to be considerably less severe.

The change-over from working shod to working unshod must be gradual. Nature's response to the increased wear on wall and sole of an unshod foot is to grow a harder and firmer horn. This process, however, takes time and until the harder horn has developed care must be taken not to work the horse to the point at which he becomes footsore.

The feet of a horse worked unshod need to receive regular attention from the farrier so that they present an even surface to the ground and so that splitting and cracking of the wall are checked.

Health, Condition, Exercise and Lungeing

HEALTH

The indications of health in a horse are as follows, and should be known to every horseman:

- Alert with ears pricking to and fro.
- Coat glossy and lying flat.
- Skin loose and supple and moving easily over the underlying bones.
- At rest, no visible signs of sweating except in very hot weather.
- Eyes open and bright, with the membranes under the lids and the linings of the nostrils salmon pink in colour.
- Eating well and chewing normally.
- Body well filled out but not gross.
- Limbs free from swellings or heat (smooth and cool to the touch).
- Standing evenly on all four feet. Resting a hind (but not a fore) foot is quite normal.

- Sound, and when walking taking strides of equal length with the weight distributed evenly over all four legs.
- Urine fairly thick, either colourless or light yellow and passed several times a day.
- Droppings, which will vary in colour with the diet, passed approximately eight times a day in the form of damp balls that break on hitting the ground. Their smell should not be offensive. When the horse is at grass, they may be rather looser but should not be as sloppy as a cow's.
- Breathing when at rest at a rate of 8–12 inhalations per minute—most easily measured by watching the flanks from behind.
- Temperature: 38°C (100–101°F).
- Pulse: 36–42 beats to the minute.

CONDITION

A horse is said to be in soft condition when his muscles are slack, when he is fat, has a gross belly and is incapable of sustained effort without sweating and distress. Unexercised horses at grass in summer are in soft condition.

A horse is said to be in hard condition when he is free of superfluous fat, both internal and external, and muscles and tendons are toned up to withstand sustained effort without injury or distress. A hunter in regular work in winter is in hard condition.

Fatness versus Fitness

It is necessary to have a clear conception of the difference between these two, and of its significance. The fat, round, sleek appearance of a horse at grass, though indicative of perfect health, is no criterion of its physical fitness to undertake work.

Fatness The condition of a horse whose muscles are soft and flabby and in an unfit state to withstand physical exertion. This applies to all muscles of the heart and lungs just as much as those of the limbs. The lung space is restricted in a horse whose belly is large and occupies an undue proportion of the frame.

Fitness The condition of a horse in hard work, capable of undertaking physical exertion without detriment to his health. Such exertion may be in the form of concentrated bursts of energy as in racing, horse trials and polo; prolonged, sustained effort as in hunting and driving; or slow, steady exertion over long periods as in trekking.

The essentials in each case are the same: the muscles of the limbs should be able to stand up to the strain imposed upon them without becoming tired or breaking down; the muscles of the heart should be equal to every demand; and the muscles of the chest should permit full and free respiration. If these conditions are fulfilled the gross belly of the fat horse will have become smaller and will no longer press on the chest, thus allowing the lungs to work more effectively.

EXERCISE

To get a horse into really hard condition is a laborious process. When hard condition has been achieved, exercise is essential to maintain it. The correct assessment of the amount of exercise needed calls for common sense. A horse that has worked hard on Monday does not need two hours' exercise on Tuesday: a quiet walk round for twenty minutes to take stiffness out of joints or swelling out of legs is all that is necessary. If, however, the horse is not to work hard again until the following Monday, then exercise or work during the intervening period is essential to maintain hard condition. The amount may be about two hours daily either led or ridden, but to keep the back and girth regions hard the horse ought to be saddled and ridden regularly.

A perfect exercising track is difficult to find, but will include a long, gradual incline over grass. A long, steady uphill trot provides everything necessary to muscle up quarters and keep wind right. Some road work is also advisable to harden the horse's legs.

The value of slow, steady exercise is emphasised. Neither galloping nor fast trotting will get or keep a horse fit—indeed they may only do harm. The route chosen for exercise should be varied—an exception to the rule given elsewhere to adhere to a fixed daily routine whenever possible.

Exercise versus Work

It is important to differentiate between exercise and work. The word 'exercise' denotes the routine activity that keeps a horse healthy and fit, without causing him excessive exertion or any loss of condition. It involves anything from, say, one to three hours at a walk or steady jog, on sound and safe going. The word 'work' implies canters, gallops, schooling work and jumping, all of which might be expected to cause the horse some real effort. Exercise and work must be sensibly co-ordinated to produce a fit, trained, horse.

Exercise under conditions in which frost and snow put a stop to hunting or other riding out of doors presents a problem. One solution is to lay down a circular straw track for the purpose.

In winter the need for adequate exercise for the stabled horse is of greater importance than for the horse kept at grass. For the latter, when keep is poor, freedom of movement in a field combined with an occasional gallop round, keeps the horse partly muscled up, and his wind right. This accounts for the fact that a hunter kept at grass during the school term or in the rider's absence may be reasonably hard when wanted, if he has been given good hay and a corn feed once a day.

To Condition a Horse for Work

To convert a horse from soft to hard condition—getting him fit—the essentials are work and food. The art lies in giving him the correct amount of each and balancing one with the other.

Assuming that the horse has been at grass for a prolonged spell without work (3–4 months) and has been brought into the stable in big, soft condition, the following is the classic procedure which is based on principles which have stood the test of time—preliminary preparation followed by a steady progression through leg- and back-hardening, muscle-building and clearing the wind. At the same time the proportion of concentrated (energy-producing and muscle-building) foods relative to bulk foods should be gradually increased.

Bearing these principles in mind, you can adjust the details to suit your own particular circumstances and those of your horse, depending on:

- The time available.
- The type of horse and how long he has been at grass.
- How fit the horse was when turned out and how fit when brought in.
- Whether he has been at grass all the time or in by day and out by night, or vice versa.

- Whether he is brought in full time from the start or in stages—i.e. for a longer period each day over a week or ten days.
- Whether he has been fed concentrates while at grass . . . and so on and so on.

The permutations are countless, but the same principles apply, so you must use your common sense and take expert advice if in doubt.

Above all, make the process a gradual one if you want the horse to stand up well to his work later on.

The following programme is suggested:

- First have the horse shod and then wormed; have his teeth inspected and, if necessary, rasped; and his annual injections given. At the same time trim and tidy him up.
- Put him into daily walking work for at least 3 weeks. Start with about 20 minutes and build up to 1 or $1\frac{1}{2}$ hours.

  Watch carefully for signs of galling under the saddle or girth and make sure that the latter are clean and pliable (see Girths, page 214). The horse's back and girth areas may be sponged with salt water and spirits if there are no injuries.
- Next combine the walking with slow trotting and increase the exercise period to 2, $2\frac{1}{2}$ or even 3 hours if time permits. Really active walking, particularly uphill, is a tremendous muscle-builder and does not cause undue strain and jar to the legs.

  A certain amount of work on a hard road helps to harden the legs but if over-done, particularly with older horses, causes jar to the feet and legs. It is also very costly in shoes. Schooling on the flat and some simple jumping can be included during this stage.
- After six weeks or so, you will probably want to do some cantering to clear the wind unless you have a good hill to work on at a slower pace. Choose good going for this and don't over-do it. Starting with half a mile and increasing gradually to a mile at a good strong canter but *not* a gallop, once or twice a week should be quite adequate.

Always remember that no two animals are the same; take expert advice if in doubt. In the case of a pony, if there is no one light enough to ride him, the exercise may take the form of lungeing, or leading from another horse.

Increase the concentrated food in proportion to the work done. If properly carried out, there will be an increase in muscle at the expense of belly. The profuse lathery sweat of the 'soft' animal will give way to a slight dampness or no dampness at all.

The whole process must be gradual, and in the case of a hunter, can hardly be accomplished in less than twelve weeks.

Some form of conditioning is also necessary with a pony at grass while the owner is away if it is to be fit on his return.

Roughing-Off

By this is understood the process whereby a fit horse, on being taken out of work at the end of his season, is prepared for a rest at grass (i.e. the 'summering' of hunters). The points to note are: stop all exercise, grooming and corn feeding; feed a bran mash daily for a few days; get your farrier to remove the shoes and trim the feet.

Choose a mild day to turn the horse out and always do this early in the day so that he can inspect his field and its fences, and can find his watering point in daylight. Alternatively, you can make the process more gradual by continuing to bring the horse in at night and making use of a New Zealand rug by day when necessary.

LUNGEING

Lungeing is a controlled way of exercising a horse while not riding him. The horse moves on a circle round the person lungeing him and is controlled by a long rein attached to a lungeing cavesson on the horse's head.

Uses

As well as providing exercise, in skilled hands lungeing may also be used for training horses, for building up their muscles and, with a rider in the saddle, for improving his (the rider's) position.

Lungeing for exercise This is particularly useful in the following cases:

When an exuberant horse needs to be settled before being ridden.

When a horse has a sore or injured back.

When a person for some reason is unable to ride a particular horse, he may exercise it adequately on the lunge, providing he is sufficiently skilled.

Safety

Lungeing should take place only in a safe, enclosed area with good footing such as a school or outdoor manège. A corner of a field with the two open sides fenced in by sheep hurdles can be an adequate substitute.

Lungeing Equipment

- Brushing boots to protect the horse's legs.
- Lungeing cavesson. This has a strong noseband, with a jointed metal nose-piece, well padded all round with a metal ring in the centre and two side rings. Some cavessons have nosebands with two joints and no padding at the back; these fit under the bit as a drop noseband. Only cavessons with a padded back-strap to the noseband should be worn above the bit as a cavesson noseband.
- Lungeing rein at least 7m (23ft) long.
- Lungeing whip long enough to influence the horse (see *figure 64*).
- Saddle or roller (unless the horse has a sore back).
- Side-reins.
- Snaffle bridle.

Fitting the Equipment

All saddlery should fit the horse comfortably. For fitting of boots see page 126.

The lungeing cavesson This should be fitted over the bridle, but its padded noseband should go under the cheek-pieces of the bridle. The cavesson should fit snugly, and the jowl straps should be tight enough to prevent the cheek-piece on the outside moving too near to the eye. Take special care to fit the noseband comfortably so that no pinching occurs between the bit and the noseband.

The lungeing rein This may be made of webbing or of rope. If the former, it should have a swivel unless the centre ring of the cavesson has one. The rein is attached to the centre ring of the cavesson.

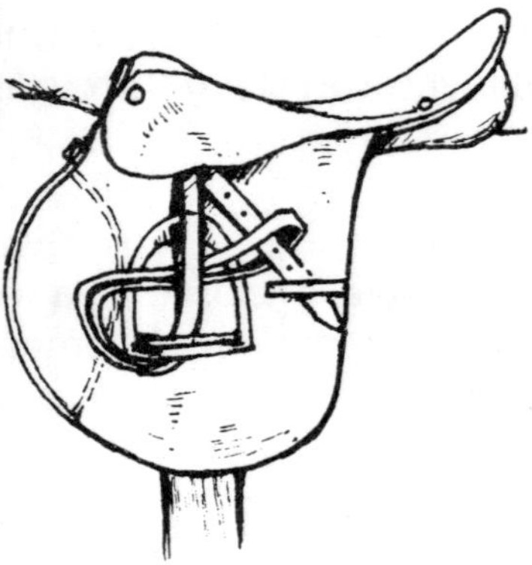

62 *A strirrup iron 'run up' and correctly secured.*

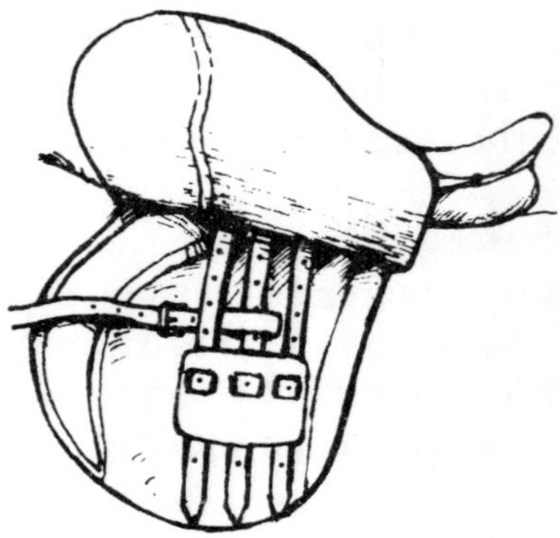

63 *A side-rein attached to a girth strap on the saddle.*

Stirrup irons If a saddle is used, 'run up' the stirrup irons and secure them *(figure 62)* or take them off and use a surcingle to hold down the saddle flaps.

Side-reins These provide additional control. Anyone intending to lunge should know how to fit and use them correctly. They run from the rings on the roller or from the girth straps on the saddle to the bit or the side rings on the cavesson. When using a saddle it helps to put the side-reins under the front girth straps and attach them to the middle ones; this prevents the side-reins from falling too low *(figure 63)*. The clip or light buckle should be fastened to the cavesson rings or to the bit, but not until you are ready to start lungeing.

The side-reins should be of equal length and long enough to permit the natural carriage, outline and movement of the horse. When not in use, they should be put over the withers and secured.

Bridle (if used) Remove the noseband from the bridle. Remove the reins or, if they are in two parts joined by a buckle, unbuckle them and put them out of the way by crossing the two halves under the horse's neck, passing them over and round, and then under the neck before re-fastening them.

Learning to Lunge

On the lunge, control of the horse is through the voice, the lungeing rein and the whip. It is helpful for anyone who has not lunged before to practise the use of all three with another person at the end of the lungeing rein. Practise until full control of the whip and rein have been achieved. Also practise the voice aids.

For teaching the horse to lunge see the Pony Club publication *Training the Young Horse and Pony*.

Method

- Wear gloves and a hard hat.
- Do not wear spurs.
- When leading the horse out to or back from the lungeing area, tuck the handle of the lungeing whip, with the lash twisted round it, under the arm furthest from the horse. This will prevent you frightening or accidentally hitting him.
- The horse will respond to the tone of voice rather than to the actual words, so make all commands clear, distinctive, consistent and meaningful.
  Use a rising tone for upward transitions and, if necessary, a slight movement with the whip to enforce the command. Use a descending tone for downward transitions and give a slight jerk to the lungeing rein if the horse continues to ignore your voice. Useful words are: 'Walk on', 'Tr-rot', 'Canter', 'Steady', 'Whoa'/'Halt', and words of praise.

When lungeing to the left Take the end of the lungeing rein in your left hand, with the slack carefully coiled or looped back and forth across the hand, and the whip in your right hand. Some people prefer to hold the slack of the rein in the same hand as the whip. Whichever method is used, the rein should be held so that it is easily made longer or shorter depending on the size of the circle required.

When lungeing to the right Take the end of the lungeing rein in the right hand and the whip in your left hand.

Do not secure your hand through the loop at the end of the lungeing rein nor allow it to become tight round your hand causing you to find yourself involuntarily tied to the horse.

The handler's position Your body should turn in the same direction as the horse is moving and should never be in front of the horse's shoulder. Generally speaking, the horse should move on a true circle, with the handler either walking in a small circle or turning on the spot *(figure 64)*.

64 *Lungeing: the position of horse and handler.*

If the whip is pointed towards the horse's girth, it will maintain the horse's impulsion, and encourage him to bend correctly on a true circle. It will also discourage the horse from 'falling in'.

- Increase the circumference of the circle as soon as possible so that the horse can trot without any undue strain. A circle which is large and easy for a 12.2hh pony will be too small for a horse.
- Insist on obedience. Try to obtain correct, active, rhythmical paces and smooth transitions. It is not necessary to canter, so do not ask for canter unless the going is good and you are confident of a smooth, correct 'strike-off' and of maintaining balance and control.
- Change the pace and the direction at least every ten minutes to prevent muscle fatigue and boredom. Halt the horse before changing direction, detaching the side-reins and putting them across his withers before turning him round. Take this opportunity to make much of him.
- The horse should halt on the circle and await your next command. If you wish to go to his head, reverse the whip so that the handle points to the front and put it under the lunge rein arm. Walk towards him, carefully coiling or looping the rein in your hand to stop it falling on the ground.

If he has not understood, walk him towards a barrier keeping level with his shoulder and give the command to halt. Very quietly point the whip at his shoulder to discourage him from turning towards you. He should then halt. When he does, walk towards him as before. Make much of him so that he knows he has done well.

- To stop the horse in an emergency, gradually make his circle smaller and then stop him either by jerking the lunge rein or by holding the whip up in front of his face.
- At the end of his exercise, walk the horse without the side-reins to cool him off.

Lungeing to improve the paces and transitions is a highly skilled art and is not covered in the above notes. However, lungeing for exercise is useful for maintaining or building muscle on the horse. It also gives a handler experience in controlling a horse from the ground.

Hunting, Competitions and Pony Club Rallies

THE STABLED, CORN-FED AND CLIPPED-OUT HORSE

PREPARATIONS

The day before Check the exact route on a map and calculate the time at which you will be setting out next morning. If hacking on, aim to journey at 6 miles an hour. If travelling by horsebox or trailer, allow for a speed of no more than 30–40 mph. If going to a meet, allow for parking and unboxing at least a mile from the meet and for hacking on from there.

If using a horsebox or trailer, check that it is bedded down with straw or shavings and that all fitments are secure, well-oiled and easy to operate; that the vehicle is serviced and fuelled; that the lighting system works on all circuits, including indicators, and that the brakes are sound.

Early morning Be up in good time and go to the stables at least an hour before breakfast.

- Look round the horse as usual; water him and tie up a small net of hay.
- Pick out feet, muck out, rearrange bedding and 'set fair' as there will be no time to do so later.
- Quarter or groom as time permits; plait the mane, put on a tail bandage or guard if travelling by box or trailer.
- Feed; remove and re-fill haynet so that it is ready on your return, and leave the horse on the short rack.
- Check saddlery—this should have been put ready the previous day.
- If using a trailer, connect the towing vehicle; ensure that the connection is secure; re-check the lighting system.
- If you have not already done so, check and load your first-aid box, haynet, water container and bucket, your grooming and saddlery bag, and such items of your personal equipment as gloves, hat, jacket, whip, etc.
- Then go to breakfast.
- Put on your own riding kit and return in good time to complete preparations for loading.

Red and green ribbons If the horse is liable to kick, you should put a red ribbon on his tail as a warning to others. This does not entitle you to ride through a crowd, particularly in a gateway, expecting other riders to keep out of your way. The responsibility for any damage by your horse—and this can be serious—is yours and yours only. If you are riding a young horse who is nervous and excitable, a green ribbon on his tail will indicate this; other riders are usually helpful in keeping clear of you, but the responsibility for your horse's actions is still yours.

Travelling by Horsebox or Trailer

Loading and unloading See detailed instructions in Transporting Horses, page 305.

On the road You may prefer to travel with your horse partly or completely unsaddled, but if when you unbox he gets excited, as is likely if other horses are passing by, it will not be easy to saddle him up. (You can, of course, put his bridle on in the box before unboxing.) However, assuming the journey is a short one it is usually more convenient to travel the horse saddled up, with a headcollar over the bridle, the reins looped round his neck or under the stirrup irons (which should be run-up). Rugs must be secured in front, or the front folded back under the roller and with a roller or surcingle over the saddle to prevent them slipping off and becoming entangled with the horse's feet. Horses generate quite a lot of heat in a trailer, particularly if there is more than one of them, so do not over-rug. In the excitement of the moment they are apt to make loose droppings which mess up the tail unless you put the bottom of it into a stocking or plastic bag, secured at the top by the tail bandage.

Parking At a meet, choose a safe place, a mile or so from the meet itself, and avoid any chance of obstructing gateways, etc. At a competition or rally, unless specifically directed elsewhere, park in a convenient place, preferably in the shade, not too close to the next vehicle, and not obstructing it or others.

When you have unboxed and completed your saddling up, store all your equipment neatly and lock the vehicle, leaving a headcollar and rope where you can easily reach them if you arrive back at the box by yourself. It is convenient to have a loop of string attached to your box or trailer to which you can tie up the horse while saddling up and on your return. But you should not tie him up if there are other horses about and yours is excitable. If hunting, hack quietly on to the meet, remembering to check your girths before setting off.

At the Meet, Competition or Pony Club Rally

On arrival If you have hacked on, dismount, look round your horse, check shoes and make necessary adjustments to your saddlery. Keep well clear of other horses' heels. If the weather is cold, re-mount and walk your horse in order to keep him warm. At a competition or rally, unbox the horse in good time to be in all respects ready and warmed up when required. Remember to say 'Good morning' to the Master, hunt staff, organiser or officials.

Once under instruction at a rally or competition, or when moved off from the meet, the task in hand must have priority, while the care of the horse, though still important, should take second place.

If the preparatory work has been well done the horse will be in hard galloping condition and equal to any demand required of him. The fact that the horse is hard, fit and keen is no excuse for galloping about without cause and jumping unnecessarily, or in any other way expending his energy needlessly. When hunting, take advantage of good going when it offers itself; at a check turn the horse's head to the wind if he is blowing hard; whenever you have the opportunity, dismount so as to ease the weight off his back for a moment; be quiet with the hands and aids; sit well *with* the horse at all paces.

Hacking back Firstly, aim at cooling off, mentally and physically. Take the quickest and easiest route. Loosen girths slightly and ride on a long rein, but do not dawdle. Your pace and route must be governed by the necessity of getting the horse into his stable cool and dry. It is the sweating, excitable horse that will cause most work when you get home. When the horse is cooled down allow him to rinse his mouth out at a stream, pond or cattle trough if one is near at hand. After the ten minute or so cool-off, jog on before you walk again. As a rule a hunter will signify when he requires to walk by gently beating his head up and down. Altering the length of your stirrup leathers also affords some relief, but slack riding is poor reward for a good horse, and now as much as ever it is incumbent on you to ride well. Even if the horse seems quite cool it is as well to walk him for the final mile or so in order to prevent the likelihood of his breaking out into a sweat once he is home. Remember that although riding into the wind appears to cool the horse, it in fact only cools the surface of his coat and he may 'break out' after he gets back to his stable. Sometimes it is advisable to dismount, slacken the girths further and lead him for the last mile or so. Your aim is to bring into the stable a calm and cool horse.

If it is raining, keep walking and jogging on. A wet, cold, tired horse takes longer to dry than a wet, warm, tired one.

Returning by box or trailer Aim to arrive back at the vehicle with the horse as cool and calm as if he were arriving back at his stable. Before loading:

- Put on a headcollar over the bridle.
- If you have no assistant to hold him, tie him up to a loop of string attached to your trailer.
- Run up the stirrup irons and *either* put on a rug over the

saddle and secure it in front and with a roller or surcingle, *or* remove the saddle and give the back a quick rub or slap to restore the circulation; then put on the rug.

• Put on a tail bandage and/or guard, and lead him into the box or trailer following the instructions for loading given on page 306.

• Once in the box or trailer, remove headcollar, take off bridle, put on headcollar again and tie him up.

• Put up his haynet.

• In the case of an unclipped horse which is still damp, use your own judgment as to whether or not to put on a rug, bearing in mind that the horse will be better off too warm than too cold.

• Before driving away make sure that the ramp is secure and the trailer legs raised; and that you haven't left any of your saddlery, clothing, etc., lying about—many a hunting whip, pair of gloves, or even a hat are lost through being left on the mudguards of a trailer.

The Road Home

The hunt, competition or rally finished, and the Master, host or organiser having been thanked, the care of the horse again becomes the first consideration. The object for both horse and rider must be to get home in a cool state, as quickly as possible, without causing any undue fatigue.

Back at the Stable

On arrival

• Dismount and run up the stirrup irons; if in a vehicle, unload.

• Look the horse over briefly, in a good light, for any obvious injuries, but *do not let him get chilled while you do this.*

• Lead the horse into the box, remove the saddle and bridle and give his back a rub and a few sharp slaps to restore circulation.

• Stand back, close the door and give the horse a chance to roll and stale.

• Put on a sweat rug and, unless very warm, another rug, secured in front and with a roller or surcingle.

- Next, give the horse a drink of water with the chill taken off—about 21°C (70°) or, if he will take it, an oatmeal or linseed gruel (see page 147) or a glucose pick-me-up if he is very tired. Many horses and particularly ponies will not drink water with anything added, so *do not delay* this drink unless you are certain that he will take the pick-me-up shortly.
- Put on a headcollar, tie up the horse and hang up a small haynet to help settle him down and get his digestion working before he has a concentrated feed.
- Loosen the headrope, sponge out the eyes and nose; re-tie the headrope and sponge out the dock.
- Unplait the mane.
- Check him over carefully for injuries—cuts, thorns, over-reaches, and treat them as necessary.
- Dry out the heels with a towel or stable-rubber. If the legs are wet and muddy, either wrap them in straw under loose stable bandages or just bandage them with thick bandages and leave them to dry—if necessary until the following morning. This has the added advantage of keeping the extremities warm.
- If the tail is wet and muddy, wash and rinse it in warm water and shake out surplus moisture.
- Thereafter your main concern is to keep the horse warm and, if he is tired, to do only enough to ensure that he is dry and comfortable. Rest and quiet will help him most.
- Keep doors and windows closed while you are working on him.
- You can do all that is necessary with the rugs over the loins or shoulders, or even by just raising one corner at a time.
- If he is still damp, put some straw under his sweat-rug, put another rug on top and leave him to dry off.
- If he is not damp, brush off the sweat marks with a body brush or glove or rough material. Remove the dry mud with a handful of dry straw or a rubber curry comb, and give him a general rub over with straw or a body brush. Pay special attention to the tender parts between the fore and hind legs and feel carefully with your hands for cuts or thorns. *Do not wash mud off* as this will almost certainly cause mud fever (see page 280). The only circumstances in which mud should be washed off are when it has caused blistering to

some part of the body; it may then be necessary to wash these parts. If so, do this gently and avoid scrubbing.

- As you proceed, keep feeling the ears since they will indicate whether the horse is likely to 'break out' or not. If they are at all cold or clammy, massage them with your hands to warm and dry them.
- When you have done what is necessary, rug him up, re-fill the water bucket and give him either a small feed or a bran mash (see page 146) and go off to your own meal.

Later that evening

- Return to the horse and feel his ears.
- If the ears are warm and dry, you have nothing further to worry about. If not, dry them off with a stable-rubber and your hands, and look for further signs of 'breaking out' over the loins and down the flanks. If these are very obvious, it may be necessary to lead the horse round to dry off, but this is risky in cold weather. Normally it is sufficient to dry the ears off with a stable rubber—remember a warm, slightly sweaty horse, is better than a cold, dry one.
- If all is well, set him fair for the night—checking rugs, ventilation, water, hay—and give him a final feed.
- If anything is amiss, keep going back until you are satisfied that the horse is comfortable.

Note

This routine is a guide only, for no two horses react in the same way, and individual conditions, stables, etc, vary enormously. In following these guidelines you should use your own good sense to achieve the result we are all after—a warm, contented, well-fed horse, settled comfortably for the night.

The day after

- Trot the horse up to see that no lameness has resulted from the previous day's activities.
- Run your hand over the saddle and girth regions to make sure that no saddle injuries are in evidence.
- Groom very thoroughly, paying extra attention to the shoes. Be especially on the look-out for girth galls, bumps, thorns and bruises, etc.

- Avoid exercise other than leading out in hand to take off stiffness and assist any minor filling out of the legs.
- Make sure that the horse can lie down and rest and has a good, deep bed.

THE GRASS-KEPT, UNCLIPPED OR PARTIALLY CLIPPED HORSE

This horse is living in a natural state at grass and therefore it is only necessary to consider what steps should be taken to compensate for any extra demands that may have been made upon him.

A horse clipped trace-high will take less out of himself by not sweating unduly, and thus keeps in fitter condition. Also he will dry off much more quickly. During the winter, horses clipped trace-high can be kept at grass providing that the clip has not been exaggerated by the removal of too much coat, or that a New Zealand rug is worn.

The evening before It is impossible to dry the thick, greasy, natural coat by rubbing, and it is also very unpleasant to sit on when it is soaking wet. So if it *is* wet, catch up the horse the evening before, pick out his feet and stable him in deep bedding to prevent draughts around his heels and legs. If his coat is very wet, and especially if it is a cold night, put on a rug with a sweat-rug underneath so that he doesn't chill while drying off. Hay up, leave water in the box and leave the top half of the stable door open, because he is used to lots of fresh air. If the horse is to have a corn feed, give this at the same time as the hay.

Early morning If the horse has been brought in over-night, look round him, replenish his water supply, pick out his feet and muck out. Brush out his mane and tail with a body brush and lay with a water brush. If the tail has been pulled (not recommended for a horse at grass), put on a tail bandage. Brush off mud and put the coat straight with the dandy brush. The natural protection against wet and cold provided by the greasy substance in the coat should not be interfered with by too much grooming. Feed the horse if he normally receives a corn feed, and then break off for breakfast.

Should the horse not have been brought in over-night, it will be necessary to fetch him from the field and wash out his feet—unless it is very cold, in which case it is better just to use the hoof-pick. Tie the horse up under cover and proceed as above. If he is very wet and muddy, any attempt to brush him over will be useless. If he is only slightly damp, it will be as well to leave the dandy brush work until after you have had your breakfast.

Setting out and on the road There is little difference in the procedure for the grass-kept horse and the stabled horse. If the former was too wet and muddy to 'quarter' properly earlier in the day, it may be possible to do so now. He will inevitably sweat when hacking on to the destination unless a quite impracticable slow pace is maintained. But steadier progress than with the stabled horse is essential, and 6–8km (4–5 miles) an hour is a fair estimate. The girths will need adjusting two or three times in the first half-hour.

On arrival The paragraph for the stabled horse applies generally, but remember that there is more strain on the lungs in fast work in an unclipped horse so care must be taken not to 'over-do' him. Excessive sweating due to his long coat will undoubtedly cause the horse to lose flesh, but nature's grease will protect him from the subsequent chilling so there is no need to be alarmed on this account providing that reasonable precautions are taken.

The road home All that applies to the stabled horse applies here, but if the unclipped horse has sweated profusely, his long coat is unlikely to dry completely on the journey home. The aim is to arrive at your destination with the horse reasonably cool under his damp coat.

On arrival home Dismount, remove the saddle and bridle and put on a headcollar. Examine his feet and body carefully to make sure that all is well. A draughty stable is a menace, so put the horse back into the field, even if he is sweating. If there are other horses in the field that are not being fed, then water and feed him before turning out. If there are no other horses in the field, turn him out straight away and then feed and hay-up in the field.

The day after Catch up the horse on a headcollar and inspect him for cuts, thorns, bruises, faulty shoes, etc. Trot him up to check that he is sound. The thorough removal of sweat-marks from the saddle region is important.

Gates

OPENING AND SHUTTING A GATE

This is comparatively simple, provided that the horse is obedient to the aids. A young horse is often anxious about going close up to a gate and finds it hard to stand still. A little time and patience are usually required.

When opening a gate, make the horse stand still, parallel with and close to it, his head facing the latch. Use the hand nearest to the gate to undo the latch; either pull or push the gate open; then pass through.

To shut the gate, turn the horse round, transfer your reins and whip to the other hand and either pull or push the gate shut. Again. make the horse stand still, parallel with and close to the gate, while you lean down to fix the latch with your hand.

If you cannot reach the latch with your hand, use your hunting crop.

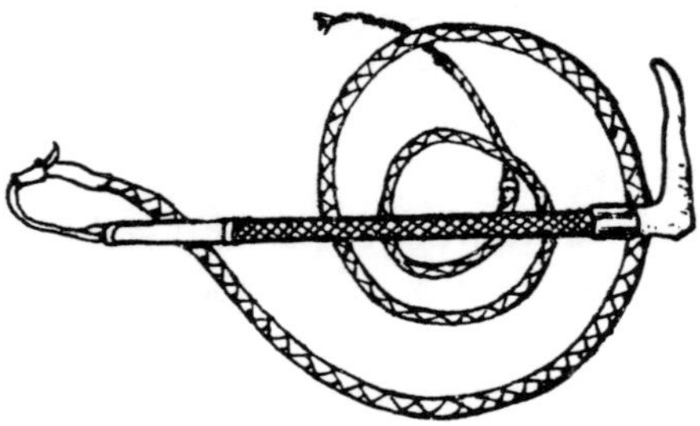

65 *A hunting crop, or whip.*

If horse and rider are not well trained to open and shut gates, much time will be lost, especially when hounds are running. When passing through a swinging gate with a group of other riders, take care not to let the gate slam in front of the person following you; push it well back so that he can catch it easily.

HUNTING

This subject is dealt with in the current Pony Club hunting publications.

Classification and Identification

BREED

For a horse to be described as belonging to a certain breed it is necessary for him to be recorded in the Stud Register of the breed society concerned or to be qualified to be so.

The following are some of the recognised breeds of light horses and ponies: *Thoroughbreds*, which figure in the General Stud Book; *Horses*: Hackneys, Cleveland Bays, Arabs and Anglo-Arabs; *Ponies*: Shetland, New Forest, Dartmoor, Exmoor, Dales, Fell, Highland, Welsh and Connemara. Stud registers exist for all of these breeds.

Half-bred

The term half-bred is used to denote a horse of whom one parent is a thoroughbred.

TYPE

Hunters, hacks, polo ponies, cobs, and vanners are *types* of horse, as distinct from breeds.

COLOUR

The deciding factor in assessing the colour of a horse or pony, lies in the colour of the 'points'. In this connection points are considered to be: the muzzle, tips of the ears, the mane, the tail, and the extremities of the four legs. White in itself is not a colour, being merely the indication of lack of colour (pigmentation).

A **black** horse is black in colour with black points.

A **brown** horse is dark brown or nearly black in colour with brown points.

A **bay** horse is a brown-coloured horse with black points.

A **chestnut** horse is a ginger or reddish colour with a similar mane and tail. 'Light', 'dark' and 'liver' chestnuts are variations.

A **grey** horse is one in which both white and black hairs

occur throughout the coat. An 'iron grey' horse is one in which the amount of black is pronounced. A 'light grey' is one in which white hairs predominate. A horse is never correctly described as a 'white horse'. A 'flea-bitten grey' is one in which the dark hairs occur in tufts.

A **dun** horse varies from mouse-colour to golden, generally has black points and shows either 'zebra' marks on the limbs or a 'list', i.e. a dark line along the back.

A **roan** horse, which may be of a 'strawberry' or 'bay' or 'blue' colour, shows a mixture of chestnut or bay and white, or black and white hairs through the coat.

A **piebald** horse is one showing large irregular patches of black and white, i.e. like a magpie.

A **skewbald** horse is one showing large irregular patches of white and any other colour except black.

Horses which conform to no fixed colour may correctly be described as 'odd-coloured'. The term 'bay-brown' is also permissible in a horse that appears to conform partly but not exactly to bay or brown.

MARKINGS

The Head

A **star** is a white mark on the forehead.

A **stripe** is a narrow white mark down the face.

A **blaze** is a broad white mark down the face which extends over the bones of the nose.

A **white face** includes the forehead, eyes, nose and part of the muzzle.

A **snip** is a white mark between the nostrils, which in some cases extends into the nostrils.

A **wall eye** is one which shows white or blue-white colouring in place of the normal colouration.

The Legs

In describing a horse, any white markings on the limbs should be defined with reference to the anatomy, and the upper limit of the marking, e.g. 'White pastern: white to fetlock or white to half cannon', etc.

The term **ermine** is employed where black spots occur on white markings.

BRANDS

Brand marks—generally indications of previous ownership—are often seen on Welsh, New Forest, Dartmoor and Exmoor pony breeds, and frequently on horses imported from America and Australia.

Brands are generally placed either on the flat of the shoulder, the saddle region or the quarters. Some animals in Britain are freeze-branded beneath the saddle.

AGE

The age of a horse is determined by reference to the front (incisor) teeth. There are six of these teeth in each jaw. During its lifetime a horse has two complete sets, namely, the *milk* (or temporary) teeth, and the *permanent* teeth. The milk tooth is small and white, with a distinct neck and a short fang. The permanent tooth is of a browner yellowish colour, much larger and has no distinct neck to it.

The change-over from milk to permanent teeth occurs at certain definite ages and the 'ageing' of a horse is based mainly upon this fact combined with the following indications:

At **one** year the horse looks young, has a fluffy tail and shows six new unworn milk teeth in each jaw.

At **two** the horse still looks young but has lost the fluffy hairs of the tail. The jaws still show a complete set of milk teeth but they are now worn.

At **three** the centre two milk teeth in each jaw have been replaced by permanent teeth which are larger and show a sharp edge.

At **four** two more milk teeth in each jaw have been replaced, namely, those lateral to the centre two. In the male a tush appears behind the corner incisor.

At **five** the corner milk teeth have been shed and new shell-like teeth show at the corners.

At **six** there is a 'full mouth' but the corner teeth have lost their shell-like appearance.

At **seven** a hook appears on the top corner tooth. A similar hook may show at thirteen years of age, and this may lead to confusion.

At **eight** the hook has disappeared, the tables of the teeth show wear, and the black hollow centres have disappeared.

From eight onwards there is no certainty as to age although, among other things, an intimate knowledge of the changes in the outline of the tables of the teeth and the slope of the jaws enables an opinion to be formed.

The age of a thoroughbred is taken from 1 January: other horses, from 1 April.

MEASUREMENT

Height

The standard measurement of height for the horse is the 'hand', which is equivalent to 10cm (4in). Shetland ponies, however, are measured in inches. In future it is likely that standard measurements will be in centimetres as well as in hands.

Measurement is made from ground level to the highest point of the withers. For accuracy, the following conditions must be fulfilled:

- The place chosen must be smooth and level.
- The horse must stand squarely on all four feet with the front feet together and the head lowered so that the poll comes in line with the withers.
- A special measuring stick provided with a spirit-level on the cross-bar should be used.

To obtain a Joint Measurement Scheme certificate, a horse must be measured without shoes, but for Pony Club competitions it is permissible to allow 12mm (½in) off the recorded height for the shoes, provided that they are normal shoes, i.e. not racing plates.

Life measurement certificates are now granted subject to two conditions: that the horse is six years of age or over; and that the measurement is taken by one of the officially appointed measurers on the panel of the Joint Measurement Scheme, with the horse standing unshod.

Height measurement has many uses. It forms part of the correct description of the horse; it provides for subdivision of horses into classes for show purposes; it is an indication of the size of a horse offered in a sale catalogue, or the size of a horse for an intending purchaser; it serves as an indication for the size of clothing, saddlery or harness on sale at a saddler.

Bone

This term is used to refer to the measurement taken around the foreleg immediately below the knee. A hunter with 'good bone' should measure 21.75cm, (8½in) or more. Where the measurement falls short of requirements the horse is said to be 'light of bone', indicating that his limbs are not up to the weight that his body should carry.

TERMS APPLIED TO HORSES AND PONIES AT VARIOUS AGES

It is not easy to define exactly the difference between a *horse* and a *pony*. It is not entirely a matter of height, though normally horses measure over 15 hands. To a very large extent the difference lies in temperament and the way in which they move. The following terms apply to both horses and ponies:

Foal At birth and up to 1 January following birth.
Colt A young male up to 3 years of age (at birth, *colt foal*).
Filly A young female up to 3 years of age (at birth, *filly foal*).
Yearling In the year after birth.
Two-year-old In the second year after birth.
Gelding A castrated male of any age.
Entire or ***stallion*** An uncastrated male.
Mare A female of any age.

A *cob* is a weight-carrying mare or gelding not exceeding 15.1 hands high with a head and neck resembling a pony, and with the body and limbs of a horse.

A *mule* results from the cross between a donkey stallion and a pony mare. A *jennet* is the product of a pony stallion out of a she-ass.

How to Describe a Horse or Pony

The following is an example of the full description of a pony: '"Kitty", a brown registered New Forest pony mare, rising five years, 13.2 hands without shoes, with a star, snip into near nostril, coronet ermine near-fore, sock off-fore, pastern partly near-hind and a stocking off-hind. Scar near-hind cannon. Branded "C.D." near saddle. Mane and tail on.'

Conformation

Conformation is the term used to refer to the physical characteristics of a horse and is often known as 'Make and Shape'. The ideal varies according to the work for which the horse is required. The points to look for, if you have the buying of a riding horse in mind, are as follows:

GENERAL IMPRESSION

Your first impression is most important:

- Is the horse attractive, with a bold, kind and generous 'outlook'; and does he appeal to you?
- Does he look intelligent and alert?
- Are the parts of his body generally in proportion?
- Does he move well? His walk will normally give a clear indication of the quality of his other paces. His action should be free, his stride of good length and straight, i.e. no 'dishing' (see page 193) or 'brushing' (see page 281).

If you are not happy so far and, in particular if you don't like the horse's expression, you should look no further.

THE POINTS TO CONSIDER

If the first impression is favourable, the following is the suggested sequence of points to be taken into account. For the sake of clarity and ease of reference, all the points are listed under each heading. In forming your general impression, you should consider only the general characteristics first—those indicated by vertical lines—and the more detailed ones later.

The Eyes

They should be large, clear and set well out at the side of the head, with width between them to give a broad range of vision and a bold, kind look. Bad-tempered horses often show the white of the eye, while those with small, deeply-set eyes ('pig eyes') may well be wilful and obstinate.

The Feet

'No foot, no horse', as the saying goes. The front feet and the hind feet should each be a pair and point straight forwards.

Avoid narrow 'boxy', donkey-like feet which, being contracted, seldom stand up to hard work.

The front feet are rounder than the hind feet and should slope at an angle of about 50° to the ground, the angle of slope of the hind feet being slightly greater. The angle of slope of each pair of feet should be the same; any difference should be regarded with suspicion.

The heels should be wide, the frog big and full to absorb concussion, while the sole should be slightly concave.

Flat feet (feet with flat soles) are inevitably weak feet; they soon get bruised and are prone to corns; the foot itself is generally larger than normal, very sloping and with low heels which do not provide any protection to the foot.

The feet should not turn outwards as this is likely to cause 'brushing'. 'Dishing', which is the result of the toe turning in, while unsightly is only serious if excessive, in which case it will cause unnecessary wear on the fetlock joint in particular and will diminish the value of the horse *(figure 66)*.

The wall of the foot should be smooth and free from rings and grooves.

The Forehand

The head This should be light, well set on, and in proportion to the size of the horse.

It should have a general appearance of leanness with width between the eyes and the branches of the jaw to allow ample room for the top of the windpipe. The angle at which the head meets the neck is most important. If it is too acute it may compress the larynx and interfere with the respiration, possibly causing a 'whistle' (see Veterinary Notes, page 298). The line of the lower jaw should be clearly defined at the angle where the head and neck meet. This means that there is enough space between them to allow the freedom of movement necessary to enable the horse to carry his head in the right position. The head should not be long and heavy as this tends to put the horse on his forehand. In profile, the line of the face should be straight and not convex i.e. 'Roman-nosed' *(figure 66)*, or concave i.e. 'dished' *(figure 66)*. 'Roman-nosed' horses are often generous and genuine, but a horse with a bump between the eyes is often wilful and stubborn.

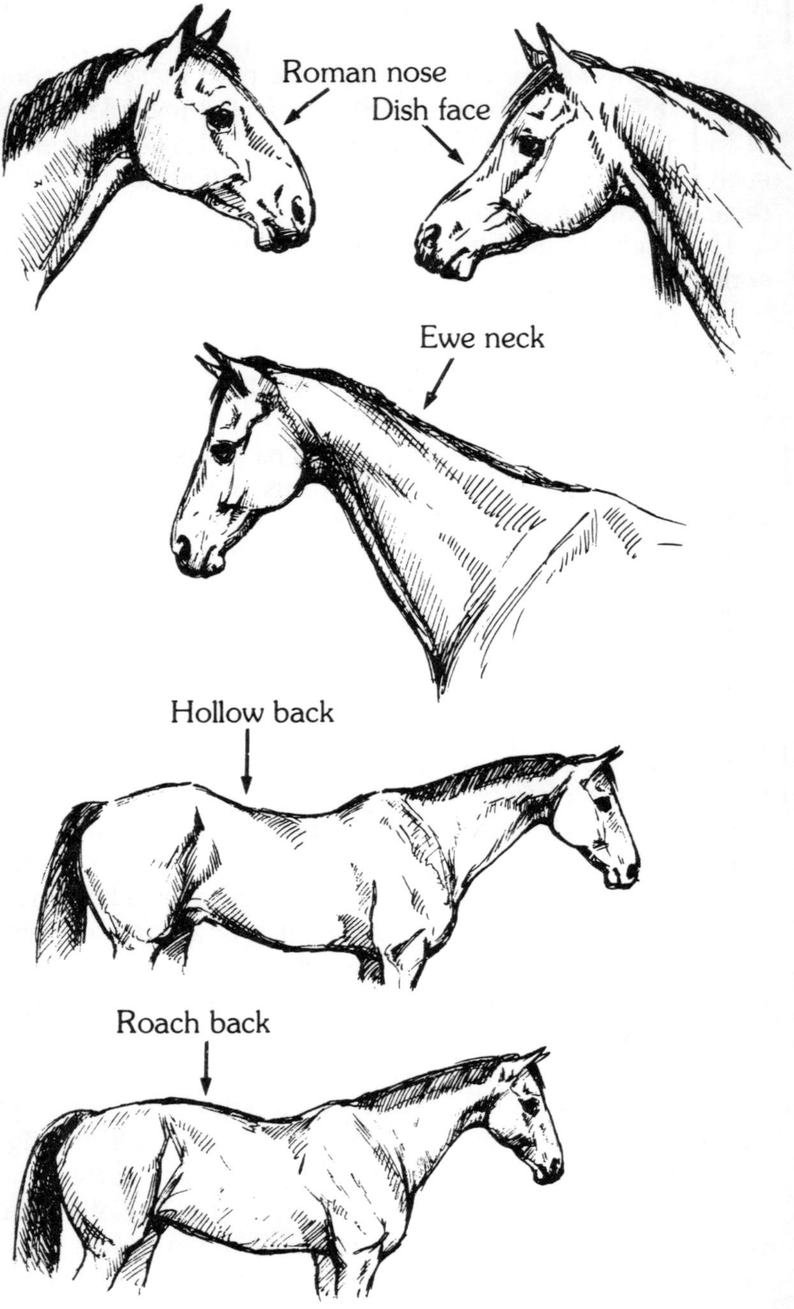

66 *Some common faults in conformation.*

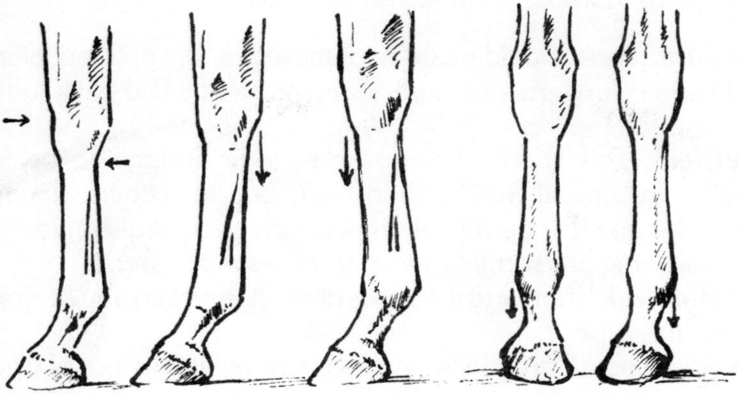

Open knee and tied in | Back at the knee | Over at the knee | Pigeon toed (toes turning in)

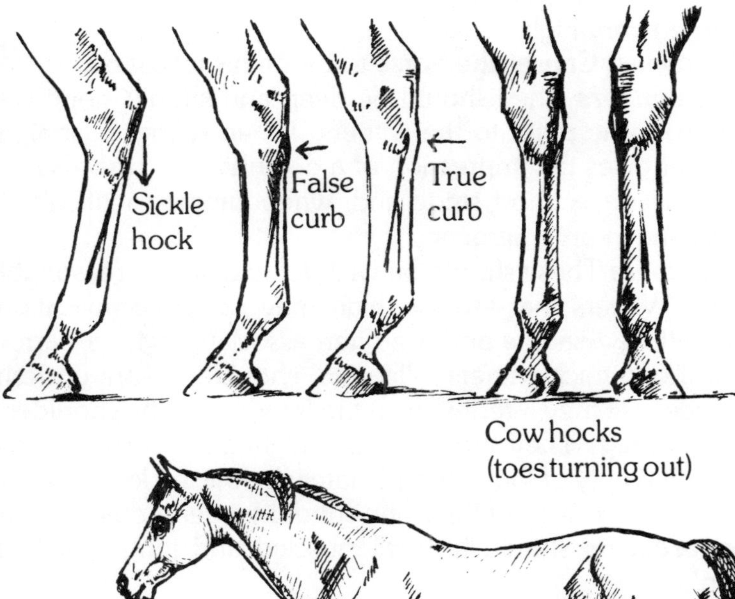

Cow hocks (toes turning out)

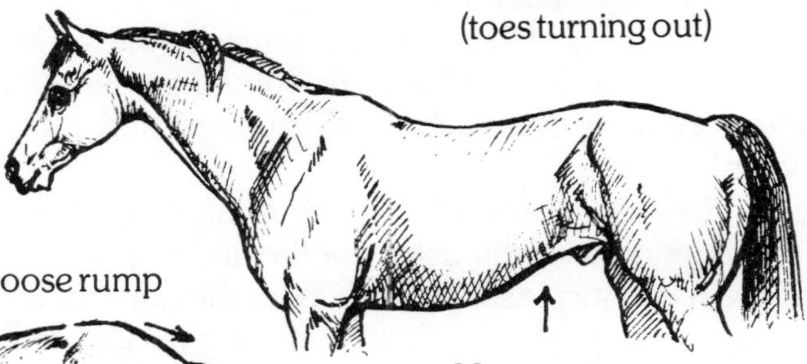

Herring gut

Goose rump

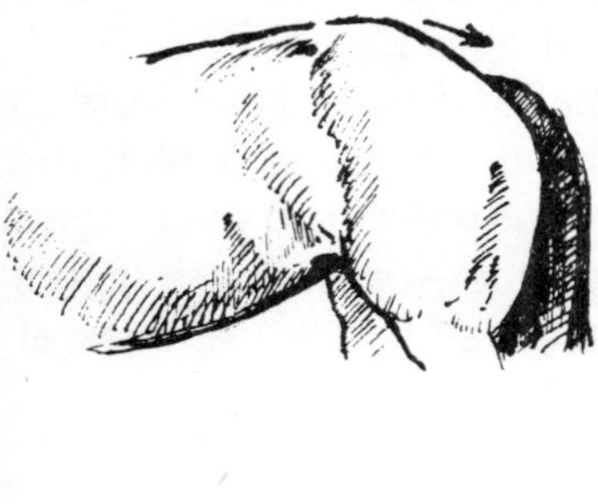

The ears These should be of medium size, finely pointed, alert and carried forward—pricked. Ears frequently laid back indicate bad temper.

The neck This should be muscular, long enough to be in proportion, and slightly arched from poll to withers. There should be no clear demarcation between neck and shoulders. The following are some types of poorly-shaped neck:

- *Bull neck* Short and thick. It is difficult to obtain any flexion with this type.
- *Ewe neck* Curved like a sheep's neck with no crest and with a bulky lower line *(figure 66)*. It causes bad head carriage, a 'poking' nose and 'star-gazing', making control more difficult for the rider.
- *Cock-throttled* Neck like a cock's, causing the head to be carried very high.
- *Low-set* Causes the horse to be on his forehand.

The shoulders They should be deep and should slope well back from the point to the withers. When riding, a 'straight' shoulder gives the impression of a downhill ride, makes for a jarring action, a short stride and, when jumping, difficulty in recovering in an emergency.

The withers These should be well-defined and of reasonable height. Withers that are *too* high may cause complications when fitting a saddle and are often associated with a narrow chest. Low, thick withers—'loaded' shoulders—are undesirable because they interfere with the freedom of the shoulder.

The front legs They should drop straight from the arm to the foot with plenty of 'bone' immediately below the knee, i.e. the circumference around the bone should be generous.

The forearms should be well-muscled and long—'well let down'.

- *The elbows* should be well clear of the body thus not interfering with the action of the limbs, which might cause strain and increase the chances of the horse injuring himself by brushing, speedy-cutting, etc, (pages 281 to 282).
- *The knees* should be broad, flat and deep from front to back, allowing enough room for the muscles which are attached to them and for the tendons which run over them.

  Indentations at the front of the knee—open knees *(figure 66)* are a sign of weakness, as are 'calf knees' i.e. back at the knee *(figure 66)*, presenting a concave rather than a convex profile when viewed from the side. These, and small apple-like knees, are often 'tied in' below the knee—the measurement round the cannon bone and tendon being smaller than that lower down—allowing only limited room for the tendon.

Knees with these defects are ill-suited to the extension required in fast work. In addition, any lateral deviation of the knee visible from the front is a serious fault, as the displacement causes strain. 'Over at the knee' or 'standing over' *(figure 66)* is the opposite of 'back at the knee'. A mild degree of this condition puts less strain on the tendons, but if exaggerated, may cause stumbling.

- *The cannon bones* should have plenty of bone, and be short and straight so that the back tendons will be less liable to strain. The tendons and ligaments should stand out in clear, hard lines.
- *The fetlocks* should have a clean appearance, be free from blemishes and signs of wear, such as windgalls, especially on the inside of the leg.
- *The pasterns* should be of medium length and slope. Long, sloping pasterns give a springy, comfortable action, but are prone to strain. Upright pasterns absorb less concussion so that more jar is transmitted to the joints higher up the leg and to the rider.

The Body

The rib-cage should be well sprung i.e. giving the barrel a rounded appearance to allow ample room for the heart, lungs and bowels and it should reach well to the back so that the loins are short.

For the same reason there should be good depth 'through the heart', i.e. the depth of the girth just behind the elbow.

The under (lower) line should slope only slightly upwards as it approaches the stifle.

Depth from the withers to the elbows, and a large girth measurement are considered indications of staying power in a horse.

A narrow-chested, flat-sided horse, with little room for the heart and lungs usually lacks stamina and is a 'poor doer'. Also the forelegs will be too close together and so the horse will be liable to brush.

If the chest is too wide, the horse will have a short, rolling gait and be an uncomfortable ride.

An under (lower) line that runs up like a greyhound's—'herring gutted' *(figure 66)* is a sign of lack of stamina and horses having this lose condition quickly with hard work.

The back A horse required for fast work should have some length of back, but the loins should be short, muscular, broad and deep, to provide a good foundation for those muscles used in galloping and jumping. Mares are sometimes longer in the back than stallions or geldings.

A horse with a long, narrow, weak back should not be considered.

A 'hollow' back *(figure 66)*, which is unduly dipped, or a 'roach' back *(figure 66)* where the curve is upward may be caused by arthritis in the vertebrae.

'Cold-backed' horses, ones which dip their backs when first mounted, should be viewed with suspicion. There is usually a reason for them doing this (spinal, muscular or kidney defect) and it is often a source of trouble.

An over-short back makes for an uncomfortable ride and the horse is apt to over-reach.

The Quarters

The quarters should be muscular with, in particular, well-muscled thighs, and plenty of length from stifle to hock (the gaskin). The hock should be large, with a prominent point and the tendons should drop straight to the fetlock.

Wide, well-muscled, flat quarters are found in fast horses; rounded quarters are typical of cobs and ponies.

Quarters that slope sharply from croup to dock often indicate good jumping ability and such horses are known as 'goose rumped' *(figure 66)*.

When seen from behind as the eye travels downwards, the quarters, in addition to being rounded at the hip, should show a gradual swell of muscle on either side, and they should be well closed up under the tail. The old adage of 'a head like a duchess and a bottom like a cook' is well worth remembering!

The points of the hips should be symmetrical and not too prominent.

The hind legs When viewed from the side, the thighs and the gaskins should appear well muscled, with the point of the hock placed directly below the point of the buttock.

The hocks As already stated these should be large, with a prominent point, wide, and deep with plenty of bone below the joint, and 'well let down'—i.e. a good length between stifle and hock.

- *'Bent' or 'sickle' hocks (figure 66)* are liable to strains such as curbs *(figure 66)* and should be avoided. To observe a curb, the horse should be viewed from the side, at right angles to the back line of the hock. A curb is a thickening of the ligament in the hind leg approximately a hand's breadth below the point of the hock and may cause a lameness when in the process of forming. A 'false curb' is a bony enlargement of the head of the splint-bone toward the outer side of the hock. Although this is unsightly, it is seldom of any consequence.
- *'Cow' hocks (toes turned out)* cause the limbs to move outwards instead of forwards in a straight line, and this results in strain *(figure 66)*.
- *'Bowed' hocks (toes turned in, hocks out)* cause the horse to twist the hock outward as the foot touches the ground. This is liable to result in thoroughpin—a soft swelling in front of the point of the hock which can sometimes be pushed through from one side of the leg to the other *(figure 87)*. Small bursal enlargements (like thorough-pins, caused by strain) are often seen on the side of the hock towards the back.
- *Bog spavins* are soft swellings on the inside and to the front of the hock. Their seriousness must be assessed by the degree of swelling and the soundness of the horse *(figure 87)*.
- *Bone spavins* are bony enlargements on the lower part of the inside of the hock *(figure 87)*. They are much more serious than bog spavins and are a permanent condition. They are best seen by standing at the side of the horse's head so that the inside of the hock is silhouetted and then each hock can be compared with the other.

When buying a horse, remember that sound movement is associated with sound conformation.

Soundness of wind is dealt with in Veterinary Notes (page 297).

Veterinary Certificate

You are strongly advised to obtain a veterinary certificate. The normal certificate is a report by the vet to the effect that he has examined the horse and found no clinical signs of disease, injury or physical abnormality that would be likely to affect its usefulness for the purpose for which you are buying it. The vet is expected to exercise the normal standard of skill of members of his profession. He is not expected to be a genius but if he misses something that a normally careful vet would have noticed he can be sued for negligence.

Few, if any, vets will nowadays give a warranty of soundness.

Safety and Insurance

SAFETY

The importance of safety has been stressed throughout this book and all horsemen must be constantly aware of the hazards involved in riding and looking after horses.

Road Safety

The dangers of riding on the highway are now so great that special instruction is essential for both horse and rider. With this in mind, the Pony Club expects members to prepare themselves and their horses to take the *Riding and Road Safety Test*. Anyone who is likely to ride or to lead a horse on the highway should read the following publications: *The British Horse Society's Ride Safely Manual; The Highway Code.*

Riding Hats and Caps

The wearing of protective headgear by Club members when mounted is obligatory at all Pony Club functions. A competitor whose hat comes off in any Pony Club competition must, on penalty of elimination, replace it before continuing. To ride without such headgear at any other time is to risk serious injury.

Riding hats, or caps, varying in quality, price and, to some extent, design, are produced by a number of hatmakers. You are strongly advised to buy one bearing the British Standards Institution's kite mark of approval and to ensure that it is a really good fit so that it remains on your head when you have a fall. If you haven't got a hat that fits really well, a chin strap will keep the hat firmly in place but these are unsightly and when worn for a long period, uncomfortable.

When buying a riding hat remember that they vary in shape as well as size, some makes being more oval than others. Both size and shape are important in achieving a good fit.

An approved pattern crash helmet with a chin strap and cover is recommended for cross-country events.

Fire Precautions

All animals are very frightened of fire, and horses are no exception. Never smoke in stables, trailers or the back of horseboxes. Where any risk of fire exists, as for example in a confined area behind the scenes at a big show, peat moss litter rather than straw should, if possible, be used.

Normally, a horse terrified by fire will not face towards it, nor leave his stable even when the doors are open.

In the event of an outbreak of fire, proceed as follows: pull the quick-release knot if the horse is tied up, remove your jacket and place it over the horse's head, his ears in the sleeve-holes, or cover his head with a wet sack. Lead the horse out of the stable; take him away from the stable area, as he may well try to return to his box, and stand him in a current of air. Inhalation of smoke is dangerous to a horse. If much smoke has been inhaled, a vet should be sent for immediately.

INSURANCE

There are a wide variety of policies available to cover most of the risks involved in keeping and using horses. An insurance broker will advise you.

2

SADDLERY

General Considerations

Saddlery, often referred to as 'tack,' is a necessary and expensive investment for horse and pony owners. It consists of the saddle and the bridle, each with their fittings, including bits, together with a variety of additional items.

There are many different kinds of each item to suit different horses and different types of riding. Here we deal only with the requirements for general purpose riding.

- It is most important that saddlery:
  Fits the horse.
  Is of the correct type.
  Is well cared for and kept clean and supple.
- Badly-fitting saddlery and hard, dry, cracking leather:
  Is dangerous for both horse and rider.
  Causes pain and injury to the horse.
  Looks sloppy and untidy.
  Devalues an expensive investment.

Well-cleaned and well-fitted saddlery adds greatly to the appearance and comfort of both horse and rider.

Buying Saddlery

Before buying saddlery it is wise to obtain expert advice from an experienced horseman or instructor.

Good saddlery needs the best leather and is therefore expensive. You can buy it second-hand, but if you do, take great care to see that:

- The stitching and leather are in good condition.
- The saddle tree is not broken (see page 209).

There are many comparatively cheap kinds of saddlery on the market—often imported—but they are unlikely to be satisfactory and can be dangerous.

Experiments are going on all the time with new materials to try to reduce the cost. Some may prove to be excellent, but meanwhile it is unwise to act as a 'guinea pig'. *Buy the best you can afford.*

Coloured saddlery should be avoided.

Bits and stirrup irons should be of best quality metal. Steel is the safest, and hand-forged stainless steel the most satisfactory. Plated metal chips and flakes off. Nickel is dangerous because it is soft, and it should not be used for stirrup irons or for bits.

There are a number of fashionable bits and gadgets which it can be tempting to buy. Many are for the use of the specialist and their merits are not discussed in this book. This is not to say that, in certain circumstances, they cannot be very helpful to you, but they should not be bought or used without expert advice.

Bits and bridles are made in standard sizes—pony, cob and full—but the size of the horse does not always govern the size of its head or width of its mouth.

In Britain standard bit measurements increase in units of $\frac{1}{2}$in, or sometimes $\frac{1}{4}$in from $4\frac{1}{2}$in, the smallest, to 6in, which is the full size. Corresponding metric sizes may not be precise equivalents.

The Saddle

A good saddle is something to be proud of and if treated with care will last a lifetime.

The purpose of a good saddle is two-fold:

- To distribute the rider's weight over the lumbar regions and the sides of the horse.
- To put the rider into the correct position.

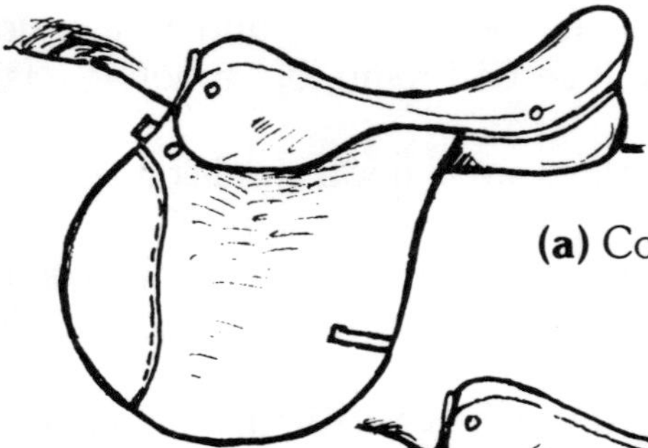

(a) Correct design.

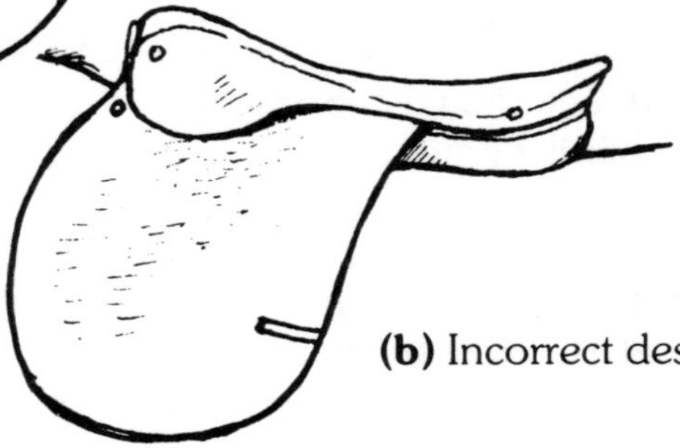

(b) Incorrect design.

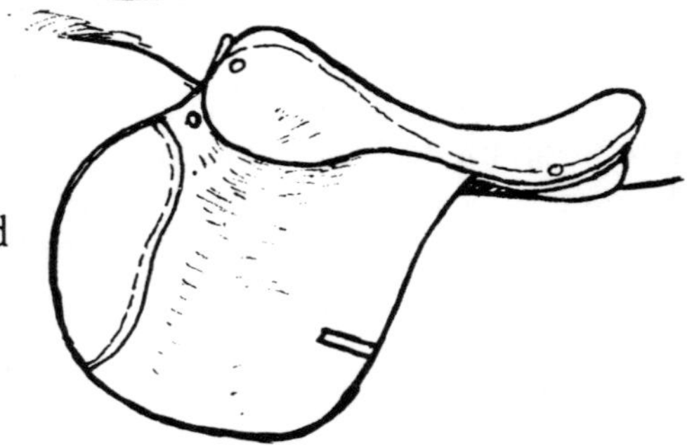

(c) A saddle in need of re-stuffing.

67 *Saddle design.*

Saddles are made in various shapes and sizes. A general purpose saddle, provided that it fits both the horse and the rider, is suitable for all Pony Club activities.

The twist, or waist, should be narrow *(figure 67a)*.

The seat must not slope to the back. The rider who sits on the back of the saddle with his lower leg stuck forward is nearly always doing so because the slope of the seat has made him slide backwards, and he is then unable to maintain the correct position of his lower leg.

A backward-sloping seat may be due to:

- Faulty design *(figure 67b)*,
- The need for re-stuffing *(figure 67c)*.

STRUCTURE

The Tree

The foundation of any saddle is the tree *(figure 68a)*. If the tree is wrong the saddle is wrong and if the saddle is wrong the rider will sit wrongly in it, however good the leather or lining.

The foundation is usually of laminated beech plywood, but various materials such as plastic and fibre-glass are also used. A spring tree saddle has a strip of flexible steel let into the tree at the waist.

Webbing

Bands of webbing, on which the stuffing and leather of the seat are carried, are stretched along the tree *(figure 68b)*.

These must be neither too tight nor too loose if the seat of the saddle is to have the right dip with its deepest part in the correct place.

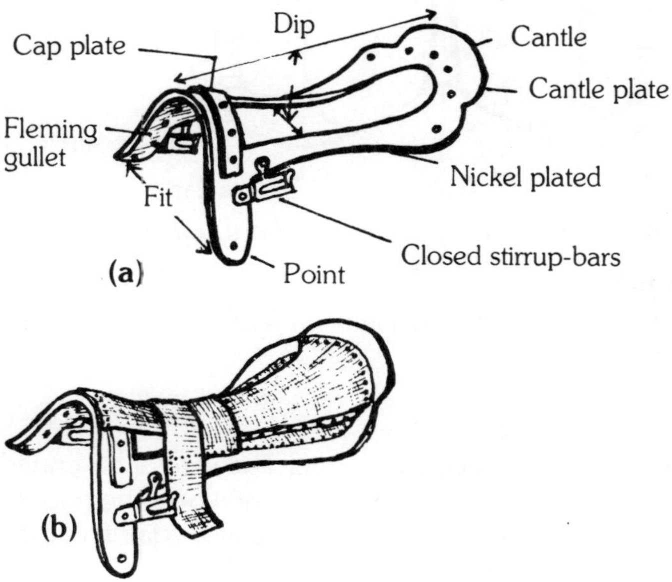

68 *The structure of a saddle.* **(a)** *The parts of a saddle tree.* **(b)** *A saddle tree after straining with bands of webbing.*

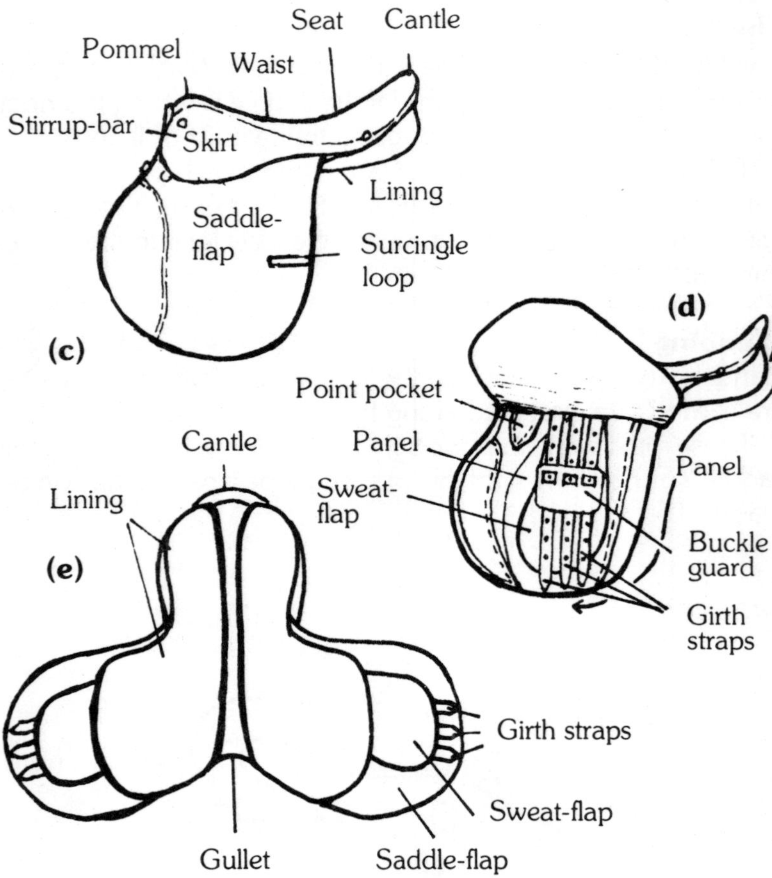

68 (c), (d) *and* **(e)** *The parts of a saddle other than the tree.*

Stuffing

This is of wool, shaped felt, foam rubber, or some other similar material.

Stirrup-Bars

These are attached to the tree. They are open-ended to allow the stirrup leathers, which are looped over them, to slip off and thus prevent a rider being dragged after a fall if his foot is caught in the stirrup iron.

On most stirrup-bars there is a hinge that allows the point to be turned up to prevent the leathers slipping off when a saddled horse is being led without a rider *(figure 68a*, page 205).

The points should never be up when the horse is being ridden.

Bars which are not open-ended or are in the form of a 'D', are not safe unless worn with safety stirrups (see page 217).

Panels

A saddle may have full or half panels which are usually stuffed with wool or shaped felt.

A full panel Reaches almost to the bottom of the saddle-flap and is lined all the way down. There is sometimes a short sweat-flap—sometimes called the 'under-flap'—between the panel and the girth straps *(figure 68d)*.

A half panel Reaches halfway down the saddle-flap and sometimes has a large sweat-flap which reaches almost to the bottom of the saddle-flap *(figure 68d)*.

Both types should be fitted with buckle-guards to prevent the panel being damaged by the girth buckles.

Linings

There are three kinds of saddle lining: linen, leather and serge.

Linen Easy to clean, dries quickly and wears much longer than serge, but not as long as leather.

Leather Easy to clean and lasts a long time if it is well looked after and the saddle used frequently.

Serge Absorbent, but does not wear well and is difficult to keep clean.

FITTING

To be absolutely perfect, a saddle needs to be fitted to the horse concerned, but this is not essential as long as the following points are observed and periodically checked:

- The weight must be evenly distributed on the lumbar muscles which cover the upper part of the ribs.
- There should be no weight on the loins; this may happen if the saddle is too long.
- Many forward-cut saddles, particularly those with a spring-tree, concentrate too much pressure on a small portion of the panel beneath the stirrup-bar, and quickly cause tenderness. A spring-tree saddle should preferably be worn with a numnah (see page 240).

- There must be no pressure on the horse's spine as this may cause serious injury. The gullet must be wide enough to prevent the stuffing bearing on the spine, and there should be enough stuffing to keep the saddle clear of the spine with the rider mounted. When first fitting a saddle and from time to time thereafter, get someone to check the back of your saddle when you are mounted to ensure that it is clear of the spine. Pressure here often goes unnoticed.
- The saddle must not pinch or bear on the horse's withers. If it is a tight fit, you need a wider tree.
- To check if the saddle is too low on the horse's withers:
  Stand in your stirrups.
  Lean forward and place your fingers between the saddle and the top of the withers.
  If your fingers are pinched, use a wither-pad as a temporary measure (see page 241). Have the tree checked and the saddle re-stuffed as necessary.
- The saddle must not hamper the movement of the horse's shoulders as it may do if the panel is too forward-cut.
- The panels must be correctly stuffed to avoid causing the rider to lose contact with his mount, and they should be regulated so as to give some support for the knee. They should be filled at the back in such a way that the rider does not slide backwards.
- The saddle must not slip forward. If it does so on a fat horse, or one with flat withers, ask your saddler to fix a 'point strap'—an extra girth strap attached to the points of the tree *(figure 68d)* and attach your girth to this and the strap next to it. Fit a crupper (see page 242).

If you are in doubt about any aspects of your saddle you must seek expert advice.

The size and width of the horse and height of the rider must also be considered, as must any peculiarities such as prominent withers. From this your saddler can decide whether a wide, medium or narrow tree or any special stuffing is required.

To Measure the Width of a Horse

It is advisable to ask your saddler to do this. Alternatively:

- Use a 1m (approximately 3ft) length of lead gas pipe or other malleable material.

- Mould it over the withers 7cm (3in) back from the shoulder where the front arch of the saddle should rest.
- Lift it off and pencil the shape (inside) on a piece of paper.

It may well be helpful to the saddler if you send this to him when you are having your saddle re-stuffed.

The size of the saddle is measured from the front of the pommel to the cantle *(figure 68c)*. Standard sizes are 38cm (15in), 40.6cm (16in) and 43cm (17in). On saddles that are cut back, it is measured from the centre of the cantle to the metal stud at the side of the pommel.

TO TEST FOR A BROKEN TREE

It is as well to know how to do this, especially when buying a second-hand saddle, but if in any doubt always seek expert advice, as a saddle with a broken tree should not be used.

The Cantle

Check that the cantle is absolutely rigid. If you can bend it backwards and forwards it means that the tree has broken.

The Waist

- Hold the saddle with both hands on the cantle and with the pommel pressed against your stomach.
- Pull the cantle gently towards yourself as if trying to make it touch the front arch.

A spring-tree saddle There should be a flexing across the centre of the waist and seat, and the saddle should feel as if about to spring back firmly into place.

If the spring-tree is weakened or broken you will feel a softness and lack of spring.

A spring-tree usually breaks more often on one side than the other, so you may well feel more 'give' on one side of the tree.

Be careful not to confuse a broken tree with a well-used spring-tree which can become very soft, but will retain its springiness.

A broken spring-tree is almost impossible to repair, as the type of reinforcement needed completely eliminates the spring effect of the tree.

A rigid-tree saddle If broken, the saddle will move noticeably on either one or both sides, but without the return spring of the spring-tree. A skilled saddler can usually repair it satisfactorily by screwing on metal reinforcing plates.

The Front Arch

If the front arch is broken, the saddle will tend to fit too low and probably to touch the top of the withers.

To check for a break:

- Place the saddle upside down on a smooth surface, preferably covered with a blanket to protect the leather.
- Place your hands on the panels on either side of the front of the saddle in the area where the points of the tree fit into the point pockets.
- Press down firmly with your hands, in effect attempting to widen the front arch. If any movement can be felt, or if there is any sound of a squeak or click from the inside of the tree, it indicates that it may be broken.
- Now examine the inside of the saddle, exerting quite hard pressure if you suspect a broken tree. You need not be afraid of breaking the tree in this way, as no human can exert enough pressure to do so.
- When a tree is in the early stages of breaking, it may also be necessary to turn the saddle the right way up and, holding the seat between the knees, push inwards on to the tree from outside the flaps.
- A saddle with a suspected broken tree should be tested both ways several times as it can be very difficult to tell whether the tree is broken or not.
- In border-line cases, very little movement may be noticed, but often you will hear a small click or squeak as two overlapping broken metal plates are moved against each other by the pressure.
- *Any indication of movement or noise* should be investigated by a saddler.
- Take care not to confuse the movement of the flexible ends of the points with that of a broken tree.
- If the tree is broken, the movement will come either from underneath the arch, which is the most usual place, or from the area of the metal stirrup-bars upwards to the arch of the saddle. If the point itself has broken below the stirrup-bar, this is not serious and it can be repaired fairly easily.

THE PAD SADDLE

Pad saddles are designed specifically for small children. They are made of felt, some being wholly or partially covered with leather. They have either no tree or just a tree fore-part.

The pad saddle with a tree fore-part or steel arch is recommended because it will sit straight on the horse and help the rider to sit correctly.

A crupper *(figure 73*, page 216) may be necessary when the pad is used on a fat pony with a round back and flat withers.

A pad saddle frequently has its own webbing girth and buckles permanently attached to it on the off side which when girthing up is then attached to a strap on the near side.

For safety, it should have two straps and buckles, or preferably, two webbing girths.

Some pad saddles, usually those without a tree, are fitted with 'Ds' instead of proper stirrup-bars. *These are dangerous*, as is any device which prevents the stirrup leather coming off the bar if pulled back. If you do use them you should have safety stirrups as well.

STANDING A SADDLE

When placed on the ground a saddle should be stood on its front arch with the girth folded so as to prevent the pommel from being scratched *(figure 69a)*. Alternatively, it can be put straight on the ground as if on a horse's back *(figure 69b)*.

A forward-cut saddle, however, will not stand on its front arch so when it is leant against a wall, for example, the girth should be folded under the front arch and up over the cantle; this will prevent them from being scratched *(figure 69c)*.

Always put the saddle down where it will not be accidentally knocked over, preferably along a fence or wall.

CARRYING A SADDLE

A saddle should be carried in one of two ways:

- With the front arch in the crook of your elbow, which allows the bridle to be carried on the same shoulder, while the other hand is free for opening doors, etc.
- Against your side, with the hand in the front arch.

These methods prevent the cantle being scraped and cut against walls and when passing through doorways *(figure 70)*.

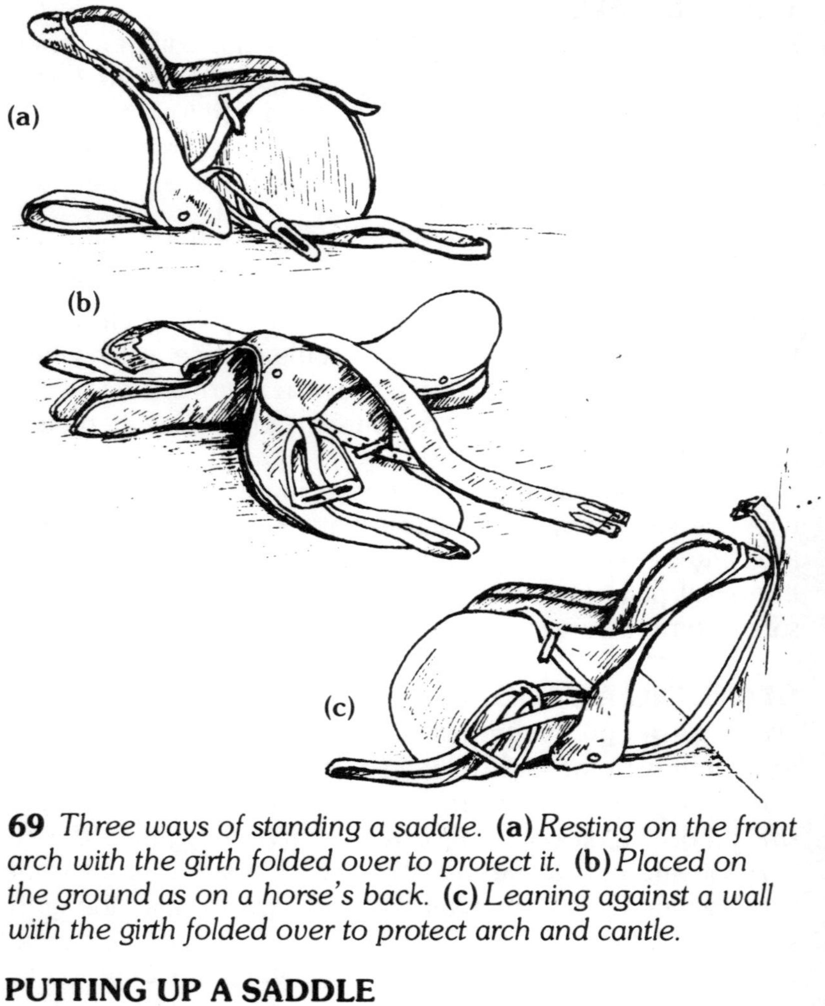

69 *Three ways of standing a saddle.* **(a)** *Resting on the front arch with the girth folded over to protect it.* **(b)** *Placed on the ground as on a horse's back.* **(c)** *Leaning against a wall with the girth folded over to protect arch and cantle.*

PUTTING UP A SADDLE

A simple and tidy way of putting up a saddle is to have a peg or bracket made for the purpose, about 46cm (18in) long, attached to the wall of the saddle-room at a convenient height:

- Place the saddle, front arch to the wall, on the peg.
- Underneath the peg have a hook from which to hang stirrup irons and four hooks fixed to the wall beside the saddle, from which the girth, stirrup leathers and martingale can be hung.
- It is better for the girth, and stirrup leathers to be put up hanging straight.

70 *Two ways of carrying a saddle.*

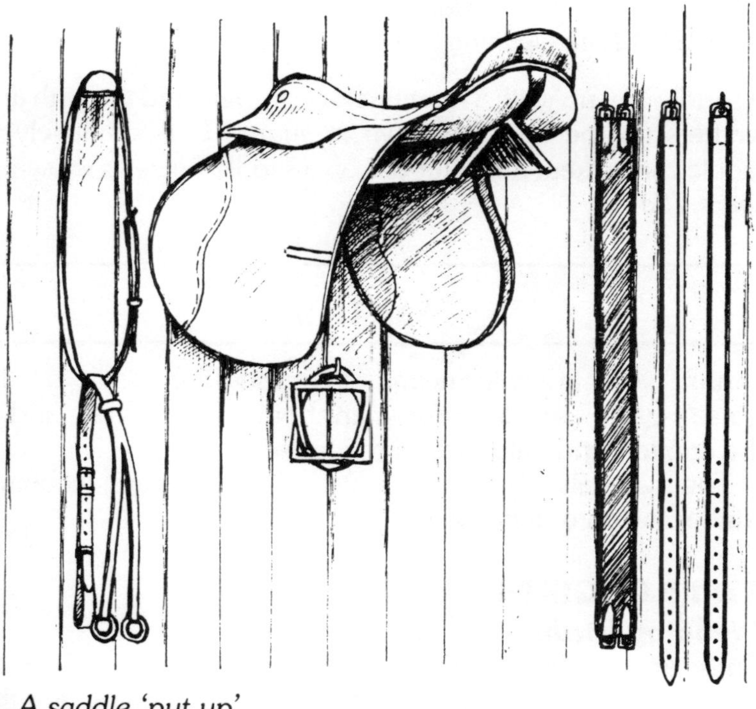

71 *A saddle 'put up'.*

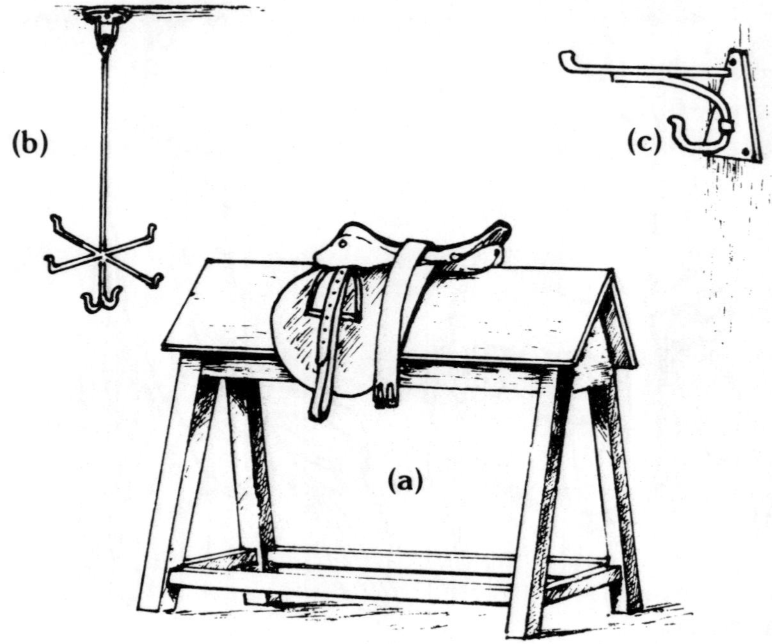

72(a) *A saddle on a saddle-horse ready for use.*
(b) *and* **(c)** *Tack-cleaning brackets.*

A saddle 'put up' with a martingale on the left, and the girth and leathers on the right is shown in *figure 71*. Alternatively, a saddle can be kept ready for use on a saddle-horse as shown in *figure 72a*.

Girths

Girths, being the sole means of securing the saddle on the horse's back, are vitally important both to your safety and to the comfort and efficiency of the horse.

The materials most commonly used for girths are: webbing, leather, nylon or lampwick.

TYPES OF GIRTH *(figure 73)*

Webbing Girths

These are cut straight with a single buckle at each end. They do not wear well and are liable to snap without warning. For this reason you should never use a single webbing girth—always use two.

Leather Girths

These are very satisfactory and wear well, but you should *pay particular attention to*:

The stitching, for your own safety.

The softness and cleanliness, for the comfort and efficiency of the horse.

There is a variety of patterns of which the following are the most widely used:

The Three-Fold A single piece of leather cut straight and folded over to form three layers with two buckles at each end. A piece of flannel or blanket soaked in neatsfoot oil should be placed between the folds to keep them soft.

The folded edge should be towards the front of the horse. If there are three girth straps, the buckles should be attached to the front two or to the first and the third.

The Balding A leather girth with two buckles at each end. The centre portion is divided into three separate straps: these are crossed over and stitched in the centre. The effect is to reduce the width of the girth where it is likely to cause girth galls. It also allows these separate straps to move on top of each other, and is thus more comfortable for the horse.

The Atherstone Made up to produce a shape similar to that of the Balding for the comfort of the horse, but consists of a piece of bag-hide folded into three layers and shaped appropriately. A leather strip is then stitched down the centre to hold the shape in position.

This girth, although easier to clean and maintain, is less popular as there is no movement of the individual parts to ensure the same comfortable fitting.

Nylon Girths

Comprised of single nylon strands joined at intervals with woven string and with two buckles at each end. They are good general purpose girths which let the air through between the strands and, on the whole, grip better than leather, particularly on an unclipped horse. For this reason they are less likely to cause galls.

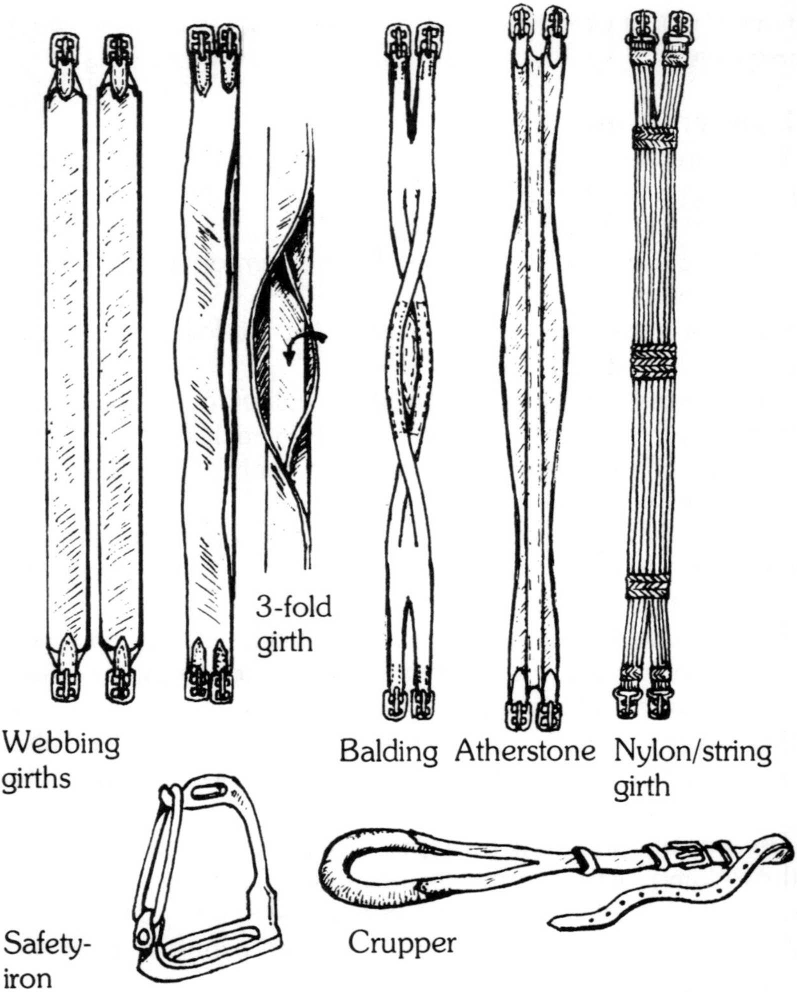

73 *Girths of various types, a crupper and a safety-iron.*

They last longer than webbing but do not last as long as leather girths.

The woven strands can become hard and 'folded', and care should be taken to keep them flat.

Lampwick Girths

Cut straight, with two buckles attached at each end. These are of very soft, comfortable, but slippery material which stretches during use and shrinks back to its former length afterwards. They need constant adjustment to prevent them from slipping while on the horse. They are normally supplied in tubular, stocking-like form with a non-stretch material such as leather or webbing inserted.

Stirrup Irons and Stirrup Leathers

STIRRUP IRONS

Irons should be of best quality metal (see page 203).

They must be large enough to allow about 1cm ($\frac{1}{2}$in) at each side of the rider's foot. The measurement should be taken at the widest part of the boot or shoe, which must have a heel to avoid the risk of the rider's foot slipping through and becoming jammed.

Nickel irons These are dangerous, being liable to break or crush.

It is particularly dangerous for small children to use adult stirrup irons which allow their whole foot to slip through.

Rubber treads are often fitted to the irons to help prevent the rider's foot slipping. They also help to insulate his feet from the coldness of the metal.

Safety Stirrups

The type most generally used has a rubber band replacing the metal on one side *(figure 73)*. The band should be on the outside, as otherwise this type of stirrup tends to make the rider's ankles turn inwards.

It is essential to use safety stirrups if the saddle is not fitted with a safety bar so that, in an emergency, the leather will release itself.

The disadvantages of these stirrups are:

- They do not hang straight due to being weighted more on one side than the other.
- The inconvenience caused when a rubber band comes off or breaks. The chances of losing the band are lessened if it is attached to the stirrup at one end by a small tight fitting loop which will retain it if the other end comes off.

STIRRUP LEATHERS

Stirrup leathers may be made of ordinary leather, rawhide or buffalo hide.

Ordinary leather When of top quality this is the smartest and most comfortable for normal riding, but it is liable to break under pressure.

Rawhide Virtually unbreakable and strongly recommended for cross-country riding. Some types are a bit thick and clumsy for dressage and showing.

Buffalo hide Reddish in colour, buffalo hide is virtually unbreakable. It is the least attractive to look at because it is usually thicker than ordinary leather or rawhide, does not tone down to match the saddle, and stretches more than other types.

All leathers stretch when new, so it is as well to wear them in at exercise before a competition, and to keep a check that, after stretching, the holes are still level with each other.

Leathers should be shortened periodically at the buckle so that the wear is not always on the same part of the leather.

The Bridle

THE PARTS OF THE SNAFFLE BRIDLE *(figure 74)*

The Head-Piece and Throat-Lash

These are made on the same piece of leather.

The head-piece in conjunction with the cheek-pieces supports the bit in the horse's mouth.

The throat-lash helps to keep the bridle in place, but when correctly fitted, i.e. not too tightly, will not prevent the bridle being pushed forward over the horse's ears.

The Brow-Band

This prevents the head-piece slipping back.

The Cheek-Pieces

These are attached at one end to the bit and, at the other, to the head-piece, supporting the bit in the horse's mouth.

The Bits

The bits most in use with this bridle are the snaffle and the

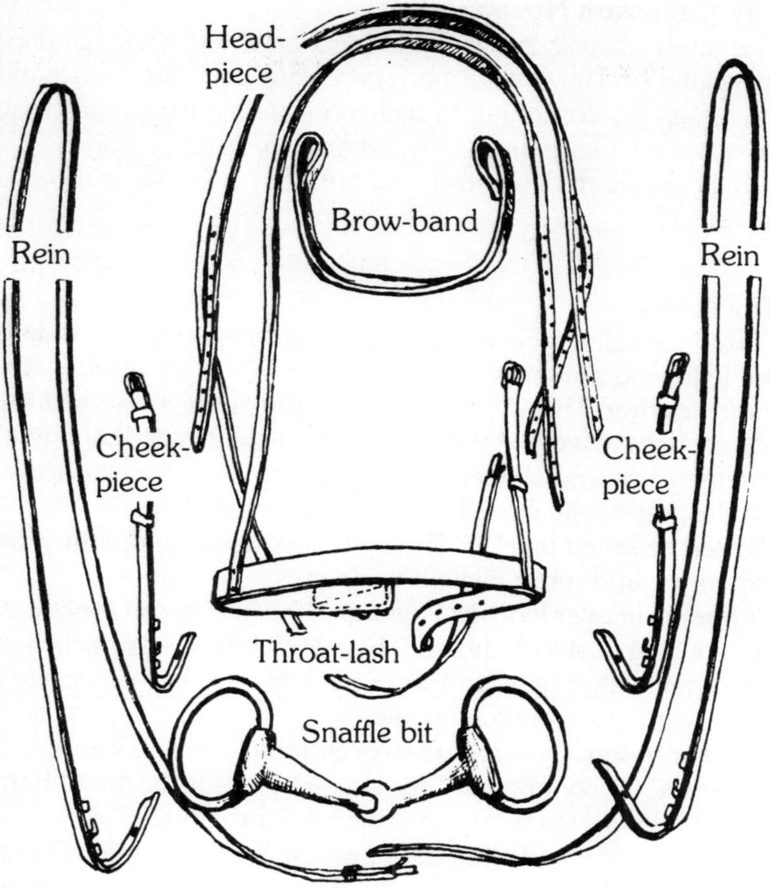

74 *The parts of a snaffle bridle.*

pelham. But a wide variety of other bits may be used (see Bits and Bitting, page 224). They are attached to the cheek-pieces and the reins by one of the following:

- Stitching, which looks best.
- Hooked billets (or buckles), if you wish the reins and cheek-pieces to be easily detachable. These are convenient as they enable you to use different bits on the same bridle. They also make cleaning easier. Buckles, although convenient, look rather clumsy.

The Cavesson Noseband

The cavesson, with its own head-piece attached, is the standard noseband and the only one, apart from the 'flash' noseband (cavesson portion only), to which a standing martingale may be attached. Other types of noseband are listed on page 235.

It is considered incorrect to turn out a horse without a noseband.

The Reins

These normally have a centre buckle and are made of one of the following materials:

Plain leather These have the best appearance and are the correct reins to wear in the show-ring and with a double bridle, but they become slippery in the rain or on a sweaty horse. Gloves, especially if made of string, overcome this.

Plaited or laced leather These are less slippery, but they are expensive and more difficult to clean.

Rubber grip over leather These give the best grip in the rain or on a pulling or sweaty horse. They do not look as attractive as the other leather reins and wear out fairly quickly, especially if worn with a running martingale.

Linen or nylon These are worn occasionally for show jumping, but are not recommended for everyday use as both are hard on the hands if used for more than a short time.

Reins are made in varying widths and should be comfortable in your hand. When for use singly they are usually wider than when for use as a pair. Apart from the show-ring where it is incorrect to wear other than leather reins, the choice between the different types is a matter of individual taste.

THE PARTS OF THE DOUBLE BRIDLE *(figure 75)*

The Head-Piece and Throat-Lash As for the Snaffle Bridle.

The Brow-Band As for the Snaffle Bridle (page 218).

The Cheek-Pieces

The cheek-pieces for the bit and for the off side of the bridoon are the same as those for the snaffle bridle. The cheek-piece for

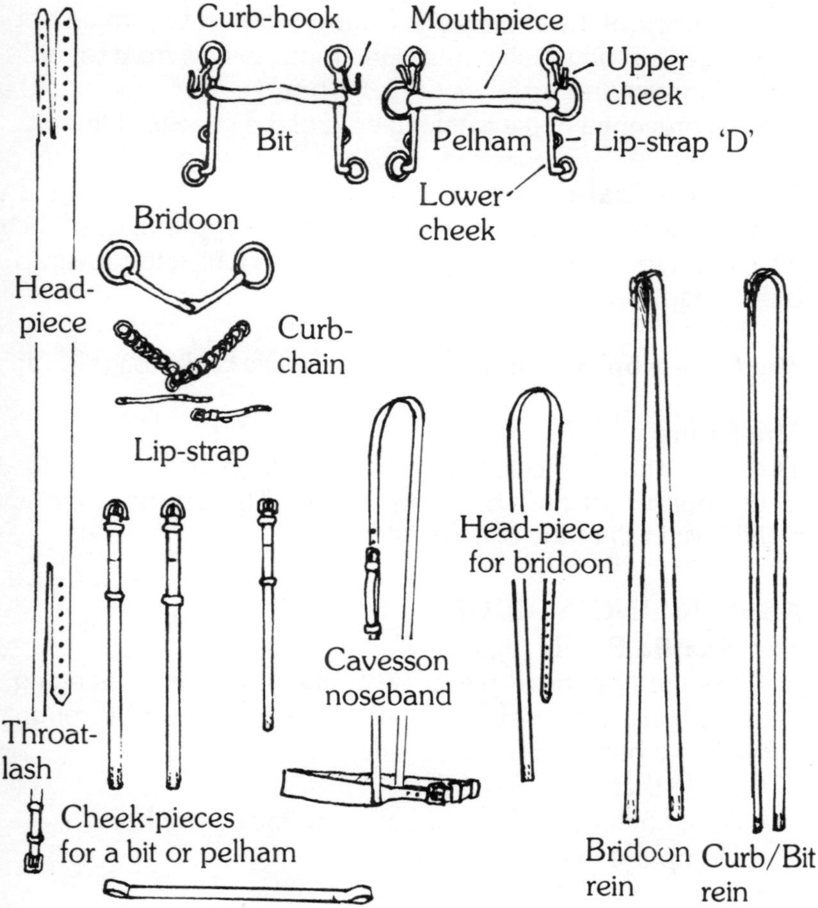

75 *The parts of a double bridle showing also a pelham bit.*

the bridoon on the near side is long, being combined with its own head-piece. It is used only with a double bridle.

The Bits (See Bits and Bitting, page 224).

The Lip-Strap

This is made in two parts: a short buckled end attached to the 'D' on the near side cheek of the bit, and a strap end threaded through the 'D' on the offside cheek. Its uses are:

- To hold the curb-chain if it becomes unhooked.

- To prevent the cheeks of a Banbury-action bit, which not being fixed can revolve round the mouthpiece, from revolving forward and up.
- To prevent a horse catching hold of the cheeks of the bit.

The Curb-Chain
This is attached to the hooks on each of the upper rings of the bit. It has a special link in the middle through which the lip-strap is threaded.

The Cavesson Noseband As for the Snaffle Bridle (page 220).

The Reins
These are attached to the rings of the bridoon and to the rings at the bottom of the cheeks of the bit. The bridoon rein is slightly wider than the curb/bit rein.

ASSEMBLING A BRIDLE
The Snaffle Bridle *(figure 74)*
- Thread the head-piece (with the throat-lash hanging behind and the noseband head-piece underneath) through the brow-band from the off side.
- Hang them on a hook.
- Assemble the bridle as if you were facing the horse.
- Do up the noseband head-piece.
- Attach the cheek-pieces.
- Attach the bit—unless already stitched on—making sure it is curving back and away from you when it is hanging up.
- Make sure that the noseband is inside the cheek-pieces.
- Attach the reins—unless already stitched on—behind the cheek-pieces.
- A hooked-billet mount is most easily put together if you thread the end of the rein through both 'keepers' first and then slide the hole back over the billet.
- Adjust all straps to their correct length and slip the ends through their 'keepers' and 'runners'. 'Keepers' are fixed loops; 'runners' are loops which slide up and down.
- If the bridle is likely to need readjustment, leave the straps out of the keepers, but slip them through the runners for neatness.

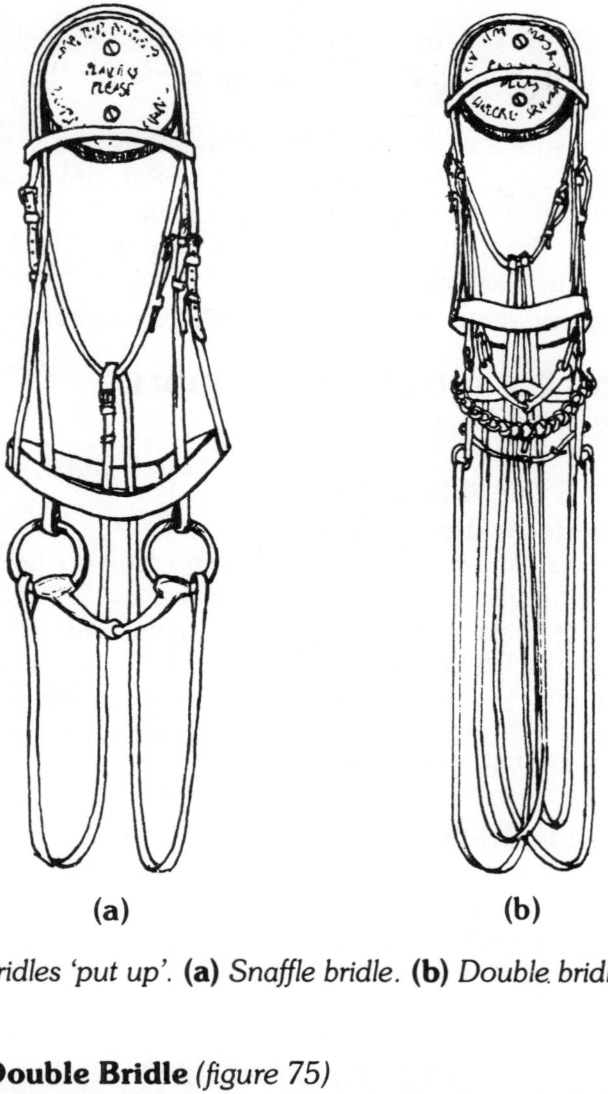

(a) (b)

76 *Bridles 'put up'.* **(a)** *Snaffle bridle.* **(b)** *Double bridle.*

The Double Bridle *(figure 75)*

- Thread the bit head-piece with the throat-lash hanging behind through the brow-band from the off side.
- Thread the bridoon head-piece underneath it from the near side.
- Thread the noseband head-piece under both from the off side.
- Hang them up on a hook.
- Assemble the bridle as if you were facing the horse.
- Do up the noseband head-piece.

- Attach the two bit cheek-pieces and the bridoon cheek-piece to its head-piece. This will give you two buckles on each side.
- Make sure that the noseband is inside the cheek-pieces.
- Attach the bit, the bridoon and the reins, unless already stitched on. The bridoon should lie above the bit.
- Attach the curb-chain and the lip-strap with the buckled end on the near side.
- Adjust all straps etc, as for the Snaffle Bridle (page 222).

PUTTING ON AND FITTING A BRIDLE

(See **Saddling Up**, page 242).

'PUTTING UP' A BRIDLE

The Snaffle Bridle *(figure 76a)*

- 'Put up' with the reins through the buckled up throat-lash and the noseband outside the cheek-pieces.
- Do not buckle the noseband, just put the end of the strap through the keeper and the runner.
- Hang the bridle on a hanger—an empty saddle soap tin nailed to the wall, for example, which will keep the head-piece in its correct shape.

The Double Bridle *(figure 76b)*

- 'Put up' in the same way as for the snaffle bridle making sure that the reins are not twisted.
- Hang the bridle up with the curb-chain hooked on in front of the bit.

Bits and Bitting

A bit is a part of the bridle that is fitted into the horse's mouth over the tongue. It is usually made of metal, vulcanite or rubber.

The principles of bitting are relatively simple unless the horse's mouth has been damaged or spoilt, or is mis-shapen due to faulty conformation.

A good rider should not need to use unorthodox bits or gadgets to achieve control.

Correct bitting, training and riding, is demonstrated by the horse's happy acceptance of the bit and quick response to any alteration of rein contact.

RESISTANCE TO THE BIT

A horse may resist the bit for any of the following reasons:

- Lack of balance and training.
- Pain or fear of the bit.
- A hard mouth.
- Temperament.
- The condition of the teeth.

Lack of Balance and Training

Before his muscles are developed and he has accustomed himself to the weight of the rider, a horse will often, through lack of balance, experience difficulty in reducing his pace. It should be realised that at this stage the horse's mouth is still 'unmade' and so, unless the rider uses tact and sympathy in reducing the pace, much harm will result. The mouth may seem hard because the horse has not yet learned to respond quickly; it will actually become hard if the rider uses force.

Pain and Fear of the Bit

A badly fitting bit or a sore mouth will often cause a horse to pull in order to get away from the pain. His natural instinct is to flee from fear of pain.

A dry mouth, a swallowed tongue, and a tongue over the bit are all evasions of pain caused by incorrect bitting and bad training.

A dry mouth is the result of the tongue being drawn back and the mouth being slightly open. The air passing through the mouth quickly dries it up. In their dry state the bars of the mouth are easily torn and bruised. A dry mouth is usually associated with a badly-trained horse.

A swallowed tongue and a tongue over the bit are caused by the horse trying to evade the contact of the bit. In either case the bars are easily damaged, and if the tongue is over the bit the under part may also be damaged. This is a serious training fault.

The conformation of a horse's head and neck is of considerable importance in bitting. If the channel which lies between the branches of the lower jaw is too narrow, or the head and neck are too closely coupled, the animal will have difficulty in flexing correctly. A horse with either of these defects is usually unpleasant to ride; if the rider persists in trying to obtain the correct head-carriage, this will cause pain and the horse, as a result, will become a 'puller'. It is better, in such cases, to allow the horse to find his own natural head carriage.

A Hard Mouth

The bit lies across the tongue and on the bars of the mouth which are extremely sensitive, being thinly covered with a skin that contains a mass of nerves. These nerves easily become numbed, then all the feeling in the mouth goes. Finally, the nerves are destroyed and the bars may develop splint-like bony lumps. When this happens, the mouth is permanently damaged and the horse termed hard-mouthed. It is a mistake to put a more severe bit on a pulling horse. The secret of a good mouth lies in the training or re-training of the horse in the snaffle, and in the horse's increased attentiveness and obedience to the rider.

Temperament

Many young horses will show excitability when taken out into the company of other horses, especially when hunting. They may throw their heads about and cease to pay attention to the rider's aids. If this happens, care must be taken not to damage the horse's mouth; ideally the horse should be ridden in a snaffle and its mouth and lips checked constantly for cuts and bruises. Any injury must be allowed to heal before the horse is ridden again. A resentful, nappy, or otherwise unwilling horse, will often display these characteristics by a refusal to accept the bit correctly.

The Condition of the Teeth

A sore mouth, although often caused by bad riding, may also be due to the condition of the horse's teeth. If they are not inspected by a vet about every six months they may become sharp and create sores on the horse's cheeks.

'Wolf' teeth may also cause great discomfort to the horse. These are small, late-developing teeth that lie in front of the top molars. They sometimes barely come through the skin. It is quite possible for the bit to touch or rub these wolf teeth, causing the horse to resist the bit in sudden irregular ways. They can easily be removed by a vet, as they are very shallow rooted. It is wise to allow the mouth to heal thoroughly before putting a bit back into the horse's mouth.

THE BITS AND THEIR USES

There are three main types of bit: the snaffle *(figure 77)*, the double bridle *(figure 78)* and the pelham *(figure 79)*. There are many variations of each. Although, unfortunately, very expensive, they are a wise investment. Cheap metals such as copper or nickel are dangerous. Bits may also be made partly of vulcanite or rubber.

The action of the bit The bit, together with other related parts of the bridle, acts on the following areas of the mouth and head:

(a) The lips and corners of the mouth
(b) The bars of the mouth
(c) The tongue
(d) The roof of the mouth
(e) The poll
(f) The chin-groove
(g) The nose

THE SNAFFLE

The young horse should first be ridden in a snaffle and his training in it continued until he is ready for a double bridle. See Beyond Basic Training, page 32.

Fitting a snaffle It is important for any bit, and particularly a jointed snaffle, to be the correct width. This can be measured by holding the bit with a hand on either side so that the joint is straight in the horse's mouth. It should protrude about 0.5cm ($\frac{1}{4}$in) on each side.

A narrow bit may pinch the corners of the mouth; if over-wide it increases the nutcracker action and makes it easy for the horse to get his tongue over the bit. If very wide it may hang down in the centre and touch the front teeth or 'saw' backwards and forwards causing soreness.

If the width of the bit is incorrect it will be impossible to make a horse comfortable in his mouth, however carefully you adjust the height.

To correct the height of the bit in the mouth, put a finger lightly on either side of the bit and press down. This is to check that the horse is not holding the bit in place himself. Then alter the bridle cheek-piece so that the bit makes the horse 'smile' but does not unduly wrinkle the corners of his lips. When the rider is mounted and has taken a contact with the reins, the bridle cheek-pieces should not sag outwards, but should lie against the horse's face. The 'drop' noseband should be fitted after the bit has been correctly adjusted.

The following are some of the different types of snaffle bit:

The Smooth-Jointed Loose-Ring Snaffle

Made of metal or rubber, with a single joint in the middle. Its action is principally on areas (a), (b) and (c) (see 'The action of the bit', page 227).

The Egg-Butt Snaffle

This has the same action as the ring snaffle, but the smooth side-joints prevent possible pinching.

The Snaffle with Cheek-Pieces

The cheek-pieces lie beside the bridle cheek-pieces and are held there by keepers. This bit will not rub the horse's mouth or pull through from one side to the other. It cannot be turned over in the mouth but lies quietly. Its action is on areas (a), (b) and (c) (see 'The action of the bit', page 227).

Some types are too heavy and clumsy for well-bred horses.

The Double-Jointed Snaffle

There are three different types:

- The double-jointed snaffle with a central piece or a plate that is rounded and lies smoothly across the tongue is sometimes called the 'French' snaffle. The action of the bit is on areas (a), (b) and (c), but the action on the tongue is mild; this often makes it a satisfactory bit on a horse with a large tongue or a narrow tongue groove.
- The third type of double-jointed snaffle is sometimes called the 'Dr Bristol'. It has a central plate, rectangular with

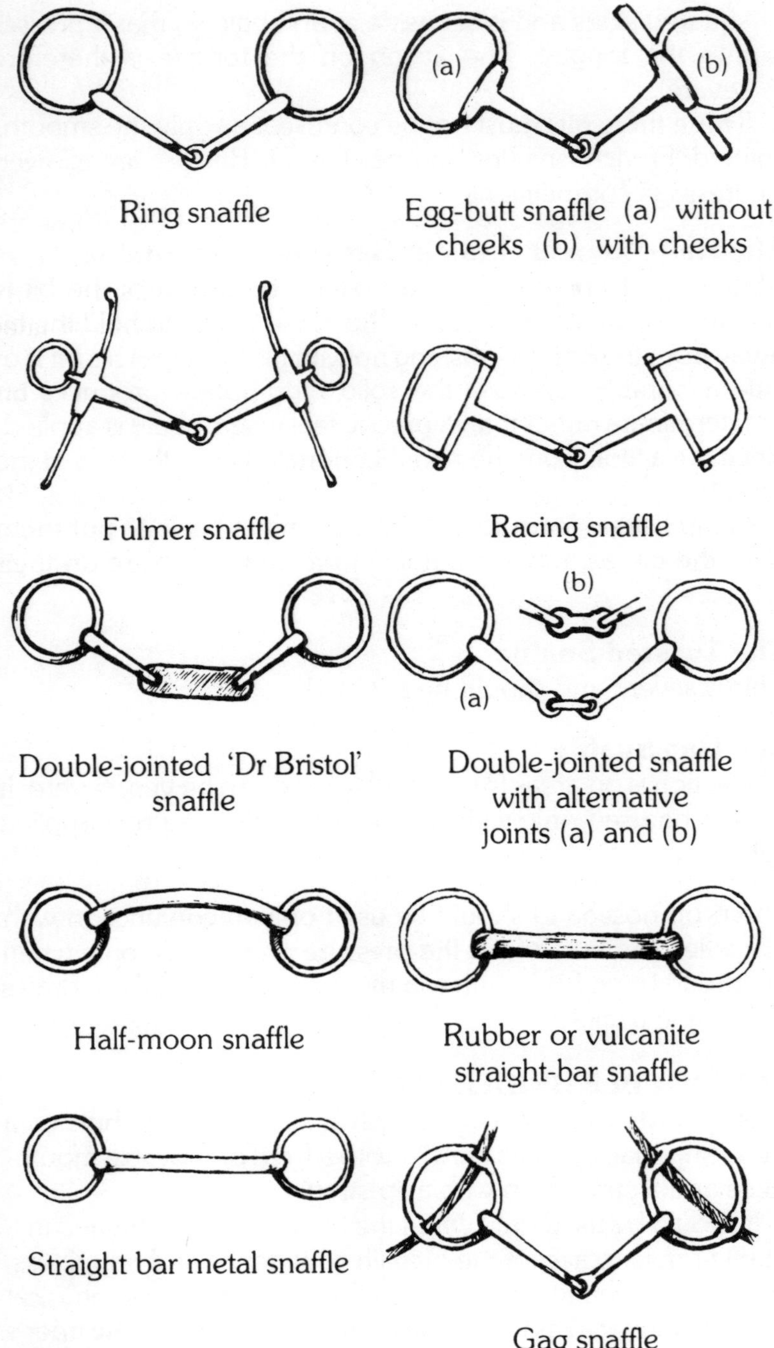

77 *Snaffle bits.*

squared sides and is set aside at an angle so that it presses into the tongue. The action on the tongue is therefore severe.

These three bits must not be confused, as only the smooth-jointed 'French' snaffles and not the 'Dr Bristol', are allowed in dressage competitions.

The Half-Moon or Straight-Bar Unjointed Snaffle

Made of vulcanite, rubber or metal. The action of the bit is primarily on areas (a) and (c). The horse is able to hold the bit away from the bars by pushing up with his tongue. This bit is of little advantage because the solid mouthpiece produces an indeterminate rather than a precise feel when an aid is applied, since the action from the rein is conducted to both sides of the mouth.

Many horses learn to draw their tongue back or put them over the bit, as they do not like to feel the pressure on their tongues.

The Twisted Snaffle

This is severe and should be avoided.

The Gag Snaffle

The action is on areas (a), (c) and (e) and can be very severe. It should be used only with two reins and the gag rein applied lightly.

The drop noseband should be used only in conjunction with a snaffle. The horse feels the pressure on his nose only when he tries to open his mouth too much, crosses his jaw or draws his tongue back.

THE DOUBLE BRIDLE BIT

This consists of a bridoon, which is either an egg-butt or a loose-ring snaffle, and a curb bit with a fixed or movable mouthpiece and a curb-chain with a lip-strap.

It should be used only when the horse has been trained in a snaffle bridle to accept the bit with confidence and happiness. He is only ready for a double bridle when his physical development makes it possible for him to respond to the rider's light aids. Only when this stage has been reached will the

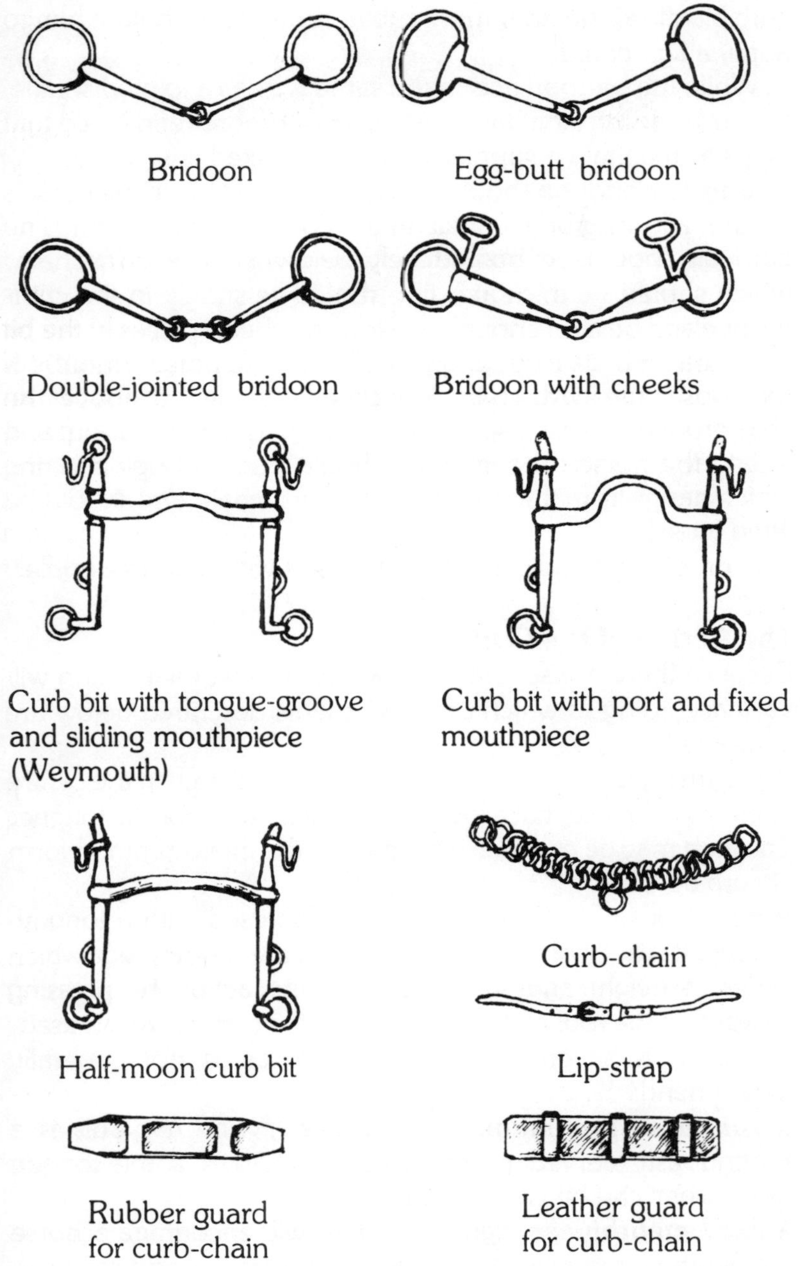

78 *Bits for a double bridle.*

curb bit afford the rider more subtle control and require him to apply a lighter aid.

While the bridoon acts in the same way as a jointed snaffle, the curb can act simultaneously to give a more refined aid that helps to maintain a supple poll and a relaxed jaw.

Fitting The bridoon should lie as high as possible in the horse's mouth, allowing one wrinkle in the corner of the mouth. The curb bit should be immediately below it. The curb-chain, which should be thick and flat, should lie snugly in the chin-groove and be tight enough to allow the cheek-pieces of the bit to be drawn back to an angle of 45° with the horse's mouth. If too loose, the curb-chain is inclined to ride up above the chin-groove when in use, or, when not in use, to flap about and irritate the horse. The lip-strap should pass through the ring which hangs from the bottom of the curb-chain and should be fitted loosely.

A drop noseband should not be used with a double bridle.

The Action of the Curb Bit

Because there are several varieties of curb bit their action will vary according to which of the elements described below are combined.

A tongue groove *(figure 78)* This is a raised arch in the centre of the mouthpiece that allows comfortable room for the tongue. The arch may be of variable height. It is the most common form of curb bit.

A port *(figure 78)* This is not to be confused with a tongue-groove. It is a 'U'-shape in the centre of the mouthpiece which varies in height and size. It comes into action by pressing forward on the roof of the mouth when the curb/bit rein is used. Its action is severe and this type of bit is not generally recommended.

A raised or half-moon mouthpiece *(figure 78)* This is a continuously curved mouthpiece that allows some tongue room when the bit is not in action.

A fixed mouthpiece *(figure 78)* This will encourage a horse to hold the bit quietly in his mouth. Its action is precise.

A movable mouthpiece *(figure 78)* While being slightly milder and encouraging the horse to 'champ' the bit, its action is fractionally delayed and therefore less precise.

Variable lengths of bit cheek-pieces *(figure 78)* The longer the cheek-pieces the greater the pressure on the poll and, through the curb-chain, on the chin-groove.

The curb bit generally presses on the bars and tongue while the cheek-pieces enable pressure to be put on the poll and on the chin-groove from the curb-chain.

The curb-chain can cause considerable pain if used without care or incorrectly fitted, especially if it slides up above the chin-groove.

The curb/bit rein should be applied only when required and not used continuously; it should be used with delicacy and subtlety. The curb bit should never be used without a bridoon, as its continuous action would soon deaden a horse's mouth.

THE PELHAM

This bit is a combination of the curb and bridoon on one mouthpiece, to the cheeks of which are attached the bridoon and curb-reins, thus trying to make the one bit perform the duties of two. This, in principle, is not a sound policy, but the fact remains that some horses will go better in a pelham than they will in anything else *(figure 79)*.

Sometimes with a pelham, a single rein is used, attached to a leather rounding, which is itself attached to the bridoon and curb-rings on the cheek of the bit. This is not recommended for use by those who wish to take advantage of the correct action of either bit, but it has proved advantageous in some cases.

A drop noseband should never be used with a pelham.

The Action

The top rein acts in the same way as a half-moon or straight-bar snaffle (page 230). The curb/bit rein acts on the chin-groove and poll *(figure 80)*.

THE HACKAMORE

The hackamore is not a bit, but it is used in place of one; it has no mouthpiece as such, but acts by leverage on the nose, poll and chin-groove. Although the mouth area is not affected by the hackamore it should be realised that the nose, poll and chin-groove are all delicate, highly sensitive areas, and just as subject to damage.

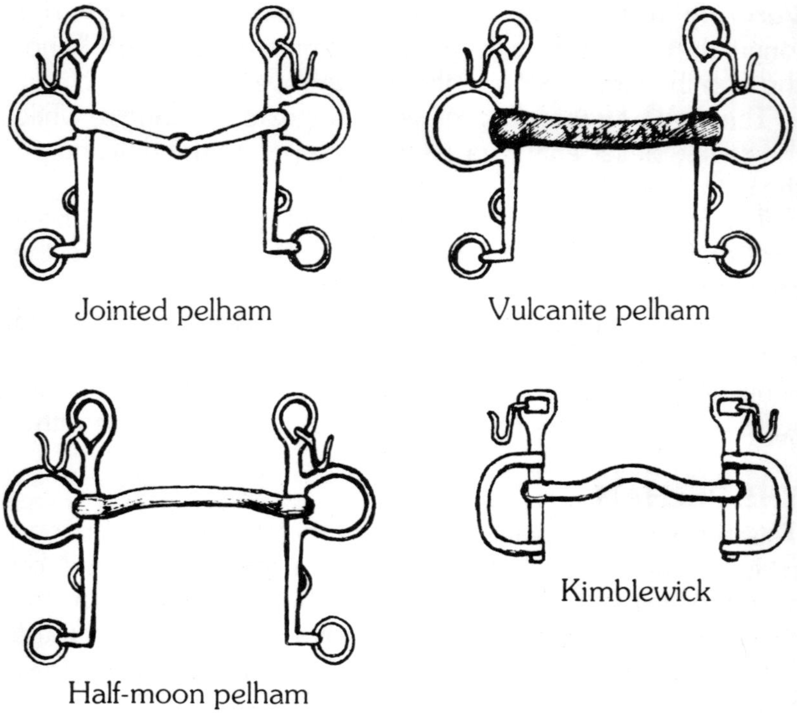

79 *Pelham bits.*

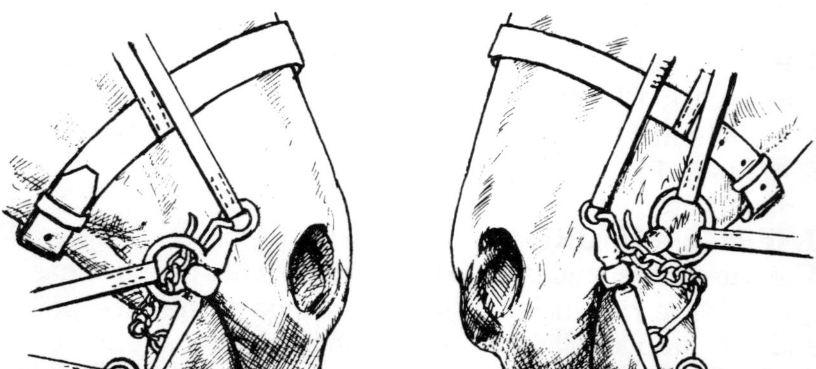

80 *Detail of the pelham and double bridle bits.*

Severe injuries that are sometimes permanent can frequently be observed after the prolonged use of this bitless bridle.

In the hands of an expert it is sometimes effective with a horse that has a spoiled, damaged or otherwise difficult mouth that will not accept a normal bit.

The hackamore does not belong in the realms of classical riding and cannot be recommended for normal use.

Additional Saddlery

NOSEBANDS

The principal types of noseband, apart from the cavesson, are:

Drop Noseband (with own head-piece) *(figure 81b)*

This is designed to prevent the horse opening its mouth wide or crossing his jaw. It is rather narrower than the cavesson and is worn below the bit.

Fitting This noseband must be very carefully fitted to avoid interfering with the horse's breathing:

- The front should be four fingers' width above the nostrils and the back should rest in the chin-groove.
- The front must remain high. This is achieved either by its first being pushed on to a spike on the ring to which it is then sewn, or by attaching a short piece of leather to the cheek-piece.
- The front piece must be of a length that allows the cheek-pieces to be well in front of the line of the horse's lips to avoid pinching or rubbing the corners of the mouth.
- It should be so adjusted that it is tight enough to prevent the horse crossing his jaw or opening his mouth wide, but not so tight as to prevent him flexing his jaw.

A snaffle bridle and drop noseband correctly fitted are shown in *figure 84* (page 251).

The Grakle (or Crossed) Noseband *(figure 81d)*

This is used for the same purpose as the drop but having an additional strap above the bit, acts over a wider area and is thus more effective in preventing the horse from crossing his jaw.

It is made up of two leather straps which cross over on the bridge of the nose and are usually rivetted together at this point. The straps pass round the horse's head and are buckled above and below the bit.

Some horses take more kindly to this type of noseband because the positioning of the straps seems to make it more comfortable.

Fitting

- Tightness should be adjusted as for the drop.
- It is important that an adjustable length of leather is used to link the top and bottom bands to prevent the bottom one slipping off the horse's chin.

The Flash Noseband

A cavesson to which a drop noseband is attached. Correctly fitted, this ensures that the drop also fits well and enables a standing martingale to be used in conjunction with it, attached to the cavesson and not to the drop.

The Kineton Noseband *(figure 81a)*

This has an adjustable leather front fitted to a metal loop on each side. The loops are hooked round the mouthpiece of the bit, between the bit-ring and the horse's mouth.

When correctly adjusted, the effect is that the pull from the reins, instead of being only on to the bit, is partially or wholly transferred to the front of the nose. This is often effective on a hard puller.

Fitting

- The front should be four fingers' width above the nostrils.
- The tighter the strap, the greater the pressure on the nose in relation to that on the bars of the mouth.
- Success lies in finding the correct balance of pressure between them.

MARTINGALES

Standing Martingale *(figure 81e)*

This consists of a strap attached at one end to the girth and at the other to a cavesson noseband, passing between the fore-legs and through a loop on the neck strap which supports it. A thick rubber ring holds the two together.

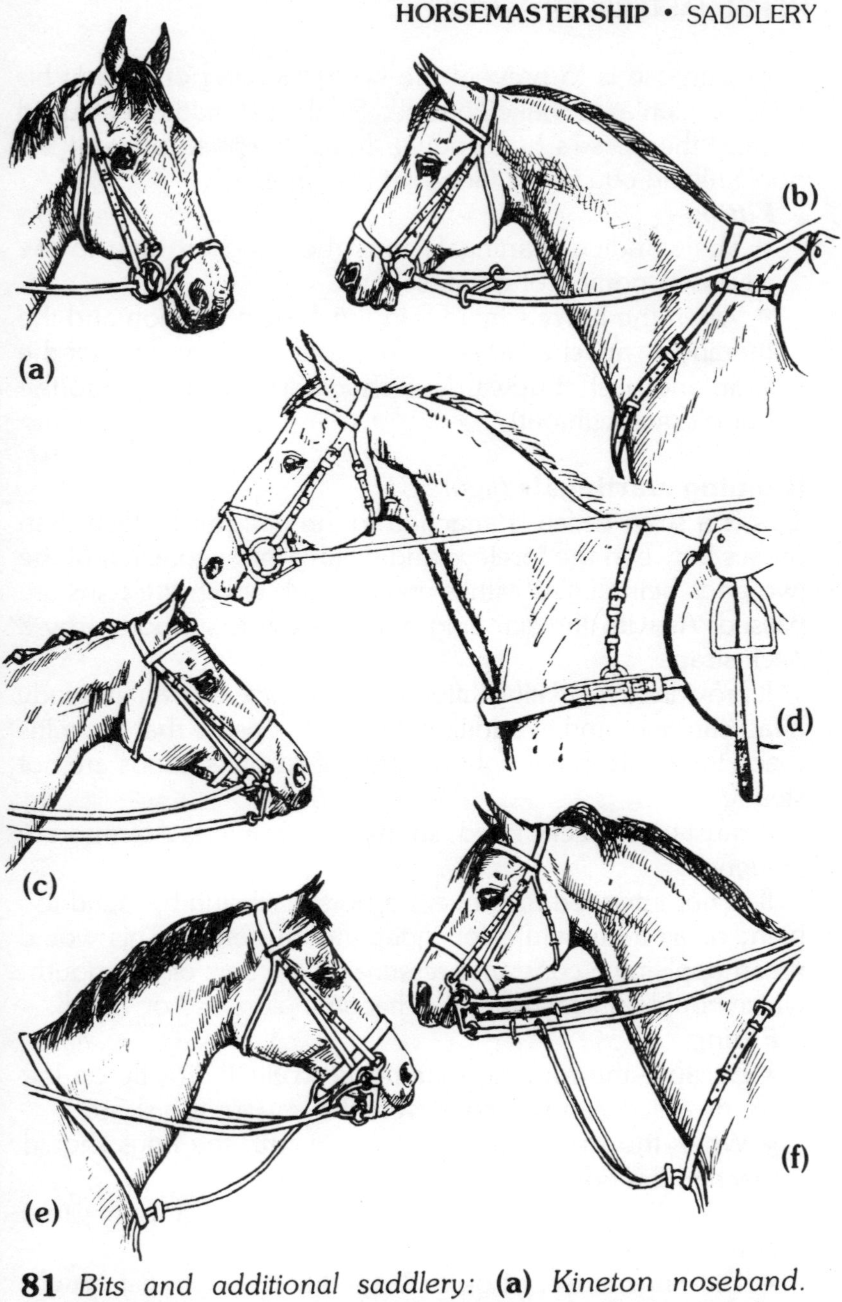

81 *Bits and additional saddlery:* **(a)** *Kineton noseband.* **(b)** *Ring snaffle; drop noseband; Irish martingale; breast-plate.* **(c)** *Double bridle; cavesson noseband.* **(d)** *Egg-butt snaffle; grakle noseband; breast-girth.* **(e)** *Pelham with roundings; standing martingale.* **(f)** *Pelham; running martingale.*

Its purpose is to prevent the horse carrying or tossing his head so high as to make control difficult. It should not be used to hold the horse's head down. It also helps to prevent the horse hitting you in the face with his head.

Fitting

- The standing martingale must be attached only to the cavesson portion of the noseband.
- When the horse's head is in the correct position and the martingale attached at both ends, put your hand under the strap and push it upwards. It should just reach up into the horse's gullet (throat).

Running Martingale *(figure 81f)*

This is a strap which is attached to the girth and which then passes between the forelegs and divides into two, each of the two ends being fitted with a ring through which the reins are passed. As with the standing martingale it is supported by a neck-strap.

It provides a regulating influence on a horse with an unsteady head carriage, and in addition it helps to ensure that the reins reach the bit from the right direction if a rider's hands are not steady.

It must be correctly fitted, so that it only exerts a temporary influence.

It is *not* intended to prevent a horse carrying his head too high, or as a substitute for independent hands. This would mean applying a constant pressure on the bars of the mouth, which could only result in their becoming bruised or 'dead'.

Fitting

- Because the running martingale works indirectly on the bit, its action can be very severe if it is worn too short.
- When the strap is attached to the girth the rings should reach to the horse's gullet (throat).
- It should be placed on the curb/bit-rein when used with a double bridle.
- The rings must not be so large that they can get caught over the rings of the bit.
- Reins with hooked billets or buckled mounts should have 'stops' on them between the rings and the bit to prevent the rings getting caught over the mounts.

- The under, or rough, side of the strap should be uppermost.

The neck strap for both standing and running martingales should fit so that it will admit the width of a hand at the withers. The buckle should be on the near side of the horse's neck.

Irish Martingale *(figure 81b)*
This consists of two rings connected by a strap approximately 10cm (4in) long.

Its purpose is to keep the reins in place and prevent them from going over the horse's head.

Fitting

- The snaffle reins pass through the rings underneath the horse's neck.

NECK STRAP

This is a simple leather strap that passes round the horse's neck. The neck strap of a martingale is the neatest, but a stirrup leather will do equally well.

Its object is, in an emergency, to provide the rider with something to hang on to other than the reins and thus to lessen the risk of him pulling on the horse's mouth.

It is a great help to young riders in rough, hilly country and when riding a horse with no mane.

It is essential when riding a young horse or teaching a rider or horse to jump.

To prevent it slipping forward when the horse lowers his head it can be attached by straps to the front 'D's of the saddle.

LUNGEING CAVESSON

(See notes on Lungeing, page 172).

BREAST-GIRTH AND BREAST-PLATE

Breast-Girth *(figure 81d)*
This is a web or elastic strap fitted across the breast and attached to the girth straps under the saddle flaps. A leather strap attached to it passes over the neck in front of the withers, and holds it in position.

The breast-girth must not be fitted so high as to restrict the movement of the neck or interfere with the windpipe.

Breast-Plate *(figure 81b)*

This is a leather neck strap attached to the front 'D's of the saddle by a strap at each side of the horse's neck, and to the girth by a broader strap passing between the forelegs. As the 'D's are apt to pull out under strain, some people prefer to use longer straps that pass under the saddle-flaps and round the girth-tabs above the buckles.

Its purpose is to prevent the saddle slipping back, which can easily happen on some horses particularly if they have 'run up light'.

Fitting

- The neck strap should be loose enough to admit the width of a hand between it and the withers.
- The straps joining the neck strap to the 'D's or the girth should lie flat without strain when the horse is holding his head normally.

NUMNAHS

A numnah is a pad cut in the shape of a saddle and usually made of one of the following:

Sheared sheepskin The best but expensive; being a natural fibre it absorbs sweat.

Nylon sheepskin Also expensive, but being a man-made fibre, does not absorb sweat.

Sorbo-rubber Semi-absorbent.

Sponge Absorbent.

Felt Absorbent.

There is a danger with non- and semi-absorbent materials of 'drawing' a horse's back (causing inflammation of the skin). It is advisable, where possible, to use natural fibres.

A numnah may be worn under the saddle, and is particularly recommended when using a spring-tree saddle. Its purposes are:

- To protect a sensitive back, especially when just starting work. Being supple, a numnah fits and lies flat on any back and thus causes less friction than the saddle.
- To protect the horse's back when riding for a long period in a spring-tree saddle which concentrates the pressure on to a limited area.
- To minimise temporarily the effects of a badly-stuffed or

ill-fitting saddle. If used for this purpose, it must be regarded solely as an emergency aid, and the saddle should be re-stuffed or replaced by one that fits the horse.

It is used by some riders when show jumping or in horse trials to prevent the horse's spine, when rounded over a fence, coming into contact with the gullet of the saddle.

It also serves as a useful guide when clipping, if a saddle-mark is to be retained.

Sponge, and materials such as sorbo-rubber squares are also used without being attached to the saddle, but they are clumsy, frequently brightly coloured and should not be used except when exercising.

Fitting

The numnah should be slightly larger than the saddle, so that when in place it is visible for about 1cm (1/2in) all round. It is kept in place by either:

- A leather strap on each side attached to the front of the numnah itself, with a loop on its other end through which one of the girth tabs is passed before being buckled up.
- An adjustable strap on each side, round the panel of the saddle.
- The panel of the saddle being slotted into a 'pocket' on the numnah.

Before fastening the girths, check that the front of the numnah is pushed well up into the front arch of the saddle to prevent it causing injury by pressing down and cutting into the withers.

WITHER PADS

Remember that the wither pad is an emergency aid and that if one is necessary it indicates that the saddle should be re-stuffed as soon as possible.

It is a piece of woollen or cloth material, or a folded stable rubber placed between the pommel of the saddle and the horse's withers.

It is used when the front arch of the saddle is pressing down on or too close to the horse's withers because the saddle is too wide (perhaps due to a broken tree) or is insufficiently stuffed.

If possible, use another saddle that does fit rather than a wither pad. Your saddler will tell you whether the saddle can be adjusted to fit by re-stuffing. If it is not possible for him to see it on the horse, take a pattern of the shape of the horse's back and send it with the saddle (see page 208).

CRUPPERS

A crupper is an adjustable leather strap with, at one end, a padded loop fitting under the horse's tail *(figure 73)*. The other end passes through a metal 'D' screwed into the back of the cantle of the saddle or is sewn on to a roller.

It is used to stop the saddle or roller slipping forward.

The tail loop on the cheaper versions is usually a piece of folded soft leather, but on the more expensive varieties is a hollow leather pad filled with crushed linseed which, when warmed by the horse's body heat, releases oil through the loop thus reducing the chance of its causing a sore by rubbing.

Saddling Up

PREPARATIONS

First, tie up the horse and keep him tied up, whether in or out of the stable, until you have finished putting on the saddle.

This saves time and is safer as it keeps him fairly still and prevents him getting away or just drifting off or nipping and kicking other horses. It stops him biting you and makes it more difficult for him to kick you or tread on your toes. Also if you have to leave him, it prevents him rolling with his saddle on and damaging it.

Brush your horse down to remove any dry mud or old sweat particularly where the saddle or girth will lie.

Collect together the tack you are going to use, with your stick, hat, gloves, etc. If you prefer, you can of course do this before tying up and brushing down your horse.

The Saddle

- Check that the stirrup irons and leathers are on the saddle and neatly run up *(figure 72)*.
- Attach the girth on the off side, leaving at least four holes above the buckles to take up if or when required.
- Check that the buckle guards are on.
- Fold the girth back over the seat of the saddle.

- If you are using a numnah, place it on the saddle-horse, and put the saddle on top, pulling the numnah well up into the front arch. Do up the straps or place the loop round the girth tabs. Fit as described on page 241.

The Bridle

- Undo the throat-lash, the noseband and, if a double bridle or pelham is being used, the strap end of the lip-strap from the buckled end on the near-side 'D', and the curb-chain from the near-side hook.
- Release the reins.
- If it is not the horse's usual bridle, adjust it so that it is certain to be big enough to go on the horse and have the cheek-pieces buckled, but not in their keepers or runners.
- Collect the martingale, if used.

These checks and adjustments are best done away from the horse in, say, the tack room, so that when you do get to him you have all your equipment to hand.

Carry your saddle as described on page 211 and your bridle with the top of the head-piece and the buckles of the reins either in your hand or over your arm or shoulder.

When You Get to Your Horse

- Hang your bridle up near by—do not leave it in a heap on the ground.
- Place your saddle over a fence or door, or on the ground as described on page 211.
- If the horse is rugged up, remove rugs unless there is any reason to keep them on after saddling: if so, see page 245.

Talk to the horse as you approach and saddle up from the near side if he is strange to you; otherwise make a habit of saddling up from either side.

It is usual to put the martingale on first, then the saddle, with the girths lightly done up, and finally the bridle. This allows the saddle to settle on the horse's back and, in cold weather, to warm up. It also helps to stop him blowing himself out or nipping at you, both of which habits he will acquire if you continually saddle and girth him up tightly all in one.

The nearer you keep to the horse's shoulder the less likely you are to be kicked.

PROCEDURE

The Saddle

- Untie the horse, put on the martingale with the buckle of the neck strap on the near side, and tie him up again.
- Where the saddle and girth lie, smooth down the hair with your hands making sure there is no mud or old sweat to cause a sore or gall.
- Pick up the saddle with the front arch in your left hand and with the cantle in your right, and place it lightly, but firmly, well forward on the withers.
- Slide it back into such a position that you will not be sitting either right on the horse's shoulder or on his loins. Never slide the saddle forward against the lay of the coat.
- See that the sweat-flap is down and that all is flat and smooth under the saddle-flap.
- If you are using a numnah see that it is lying flat and is pulled well up into the front arch of the saddle.
- Go quickly but quietly under or in front of the horse's neck to the off side. Let down the girth and see that all is smooth and flat under this flap also. It is very important that you do go round to do this as you cannot check properly by leaning over the top of the saddle.
- Return to the near side and, keeping your shoulder close to the horse's shoulder, bend down and take hold of the girth. Put it through the loop of the martingale and buckle it up so that, though it is not too tight, it is tight enough to prevent the saddle slipping back. As a rough guide you should be able to insert the flat of your hand between it and the horse. Be sure there is no skin pinched behind the elbows and that the buckles of the girth are level.
- Having tightened up the girth there should be at least four holes remaining on the girth-strap above the buckle on this (near) side also. If there are not, the girth is too long. Leave the buckle-guards up above the buckles until the girth is finally tightened. The tightness of the girth should be checked periodically during your ride especially in the case of a horse that 'blows himself out'. The girth should be tight enough to ensure that the saddle doesn't slip but not so tight as to cause unnecessary discomfort to the horse.

The Crupper *(figure 73)*

The crupper, if used, should be put on after the saddle is girthed up.

- Standing close to the near hind leg, gather up the tail in the right hand and pass it through the crupper.
- Take care that the crupper lies well to the top of the tail and that all the hairs are through it and lying flat.
- Pass the strap through the 'D' on the back of the saddle and adjust the length so that it steadies the saddle, but is not so short that it pulls up the tail.

To Saddle a Rugged Up Horse

Follow the procedure given in Preparations, page 242. Then:

- Untie the horse; put on the martingale, if used, with the buckle of the neck strap on the near side, and then tie him up again.
- Undo the roller and put it neatly on the manger.
- Undo the front buckle and fold the rug back over the horse's loins.
- Put on the saddle and do up the girth.
- Draw the front of the rug forward over the saddle.
- Do not buckle the front strap because if the rug slips, as it may well do with no roller to hold in on, it will drop clear instead of one end remaining caught up round the horse's neck with the other end under his feet. This will both frighten the horse and damage the rug.
- Put on the bridle and, if you are not leaving immediately, replace the headcollar over it and tie the horse up again.
- If you intend to travel the horse in a box or trailer with a rug or rugs on, refer to Clothing in Transporting Horses, page 305.
- If you are not travelling, before leading the horse out of the box strip off the rug, fold it neatly and put it on the manger or windowsill, or over the door if it is under cover. Untie the horse; remove the headcollar and either hang it up or put it neatly where it is easily accessible on your return.

Putting on and Fitting a Bridle

Two methods of putting on a bridle are shown in *figure 82*. The first steps are the same for both methods:

- Taking the bridle from where you hung it up, put your left

Method 1

Method 2

82 *Two methods of putting on a bridle.*

hand under the brow-band and head-piece, and place the bridle over your forearm with the brow-band nearest to your elbow.

- Place the buckles of the reins in front of the head-piece on your left forearm, leaving both hands free.
- Untie the horse's head rope.
- Place the reins over the horse's head and neck. This gives

you something with which to hold on to the horse when you remove the headcollar.

- Remove the headcollar or halter and hang it up—do not just drop it where you and the horse can tread on it.
- If the horse is in a box or stable, turn him to face the light, and thereafter follow either Method 1 or Method 2.

Method 1 *(Figure 82)*

- Take hold of the head-piece of the bridle with your right hand.
- With your left hand under the horse's muzzle allow the mouthpiece of the bit to rest on your first finger and thumb (*figure 82*).
- Bend your first finger and feel between the horse's lips on the off side where there is a gap between the teeth. This makes him open his mouth.
- Keeping your right hand close to his forehead, draw up the bridle, using your left hand to guide the bit gently into the horse's mouth.
- The left hand can now assist the right hand to pass the head-piece over each ear in turn. Take care to smooth the mane and forelock and run your finger round under the head-piece to be sure that nothing is twisted.
- Do up the various buckles starting at the ears and working down. Be sure that, as you do up each strap, you put the end neatly through its keeper and runner.
- Do up the throat-lash, allowing the full width of your hand between it and the side of the jaw bone, when the horse's head is up. It must never be so tight as to interfere with the horse's breathing or flexing.
- Stand in front and make sure that:

The brow-band is level—just below, but not touching, the ears, and of a length that will not interfere with the hang of the bridle by pulling it forwards.

The noseband is level.

The bridle itself is on straight.

The bit or bits are level in the horse's mouth.

- Buckle the noseband allowing two fingers' width between it and the front of the horse's face, thus allowing the horse to open his mouth slightly and to flex his jaw. Adjust the length of the noseband on the near side so that it lies midway between the projecting cheek-bones and the mouth.

Straighten it by easing up its head-piece just below the brow-band on the long side and then easing it down on the other. Don't try to pull it through both loops of the brow-band at the same time as this will almost certainly pull the whole bridle crooked. Check that it is adjusted so that its side-pieces will not touch the horse's eyes.

- When using a running or Irish martingale, unbuckle the reins—curb/bit rein when using a double or pelham bridle—pass them through the rings with their under or 'rough' side uppermost, and re-buckle them.
- Adjust the bit on both sides, counting the holes to make sure it is level. For the correct method of fitting, see Bits and Bitting, pages 227 and 232.
- Adjust the bridoon on the off side, easing it up or down as necessary as with the noseband. See also Bits and Bitting, page 232.
- ***Fit and adjust the curb-chain as follows:***

Attach one end to the off-side hook *(figure 83a)*.

The curb-chain must be made to lie flat in the chin-groove.

From the near side, twist the chain to the right until it is flat.

The loose ring in the centre should be hanging down *(figure 83b)*. If it is not, remove the end link, turn it over and replace it so that it is.

Keeping the chain twisted to the right and changing your grip so that your right thumb is to the rear and underneath, place the end link on the hook *(figure 83c)*.

Shorten the chain to the required length on the near side unless it is more than, say, 3 links too long, in which case take it up an equal amount on each side.

To shorten on the near side Take hold of the appropriate link with your right thumb at the rear and underneath and, pushing it up towards the horse's jaw, place it on the hook making sure that your thumb is still underneath *(figures 83d* and *e)*.

To shorten on the off side Follow the same procedure taking hold of an off-side link.

It is important that the links are fitted in this way because otherwise the top and not the flat edge of the links will bear

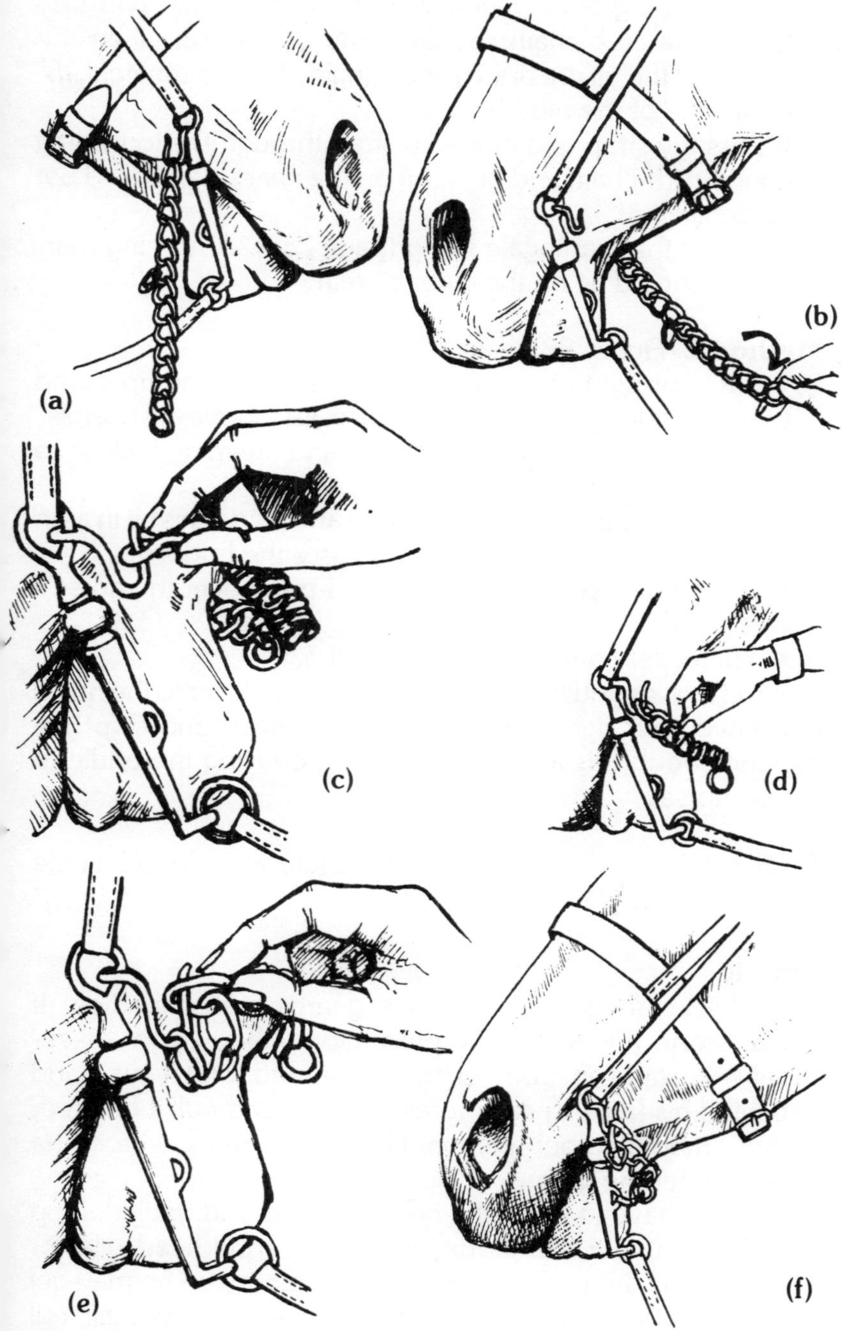

83 *The correct way to fit a curb-chain.*

on the chin groove. *Figure 83f* shows the chain correctly fitted. The curb-chain should come into action when the cheeks of the bit are drawn at an angle of approximately 45° to the line of the mouth.

- Pass the strap end of the lip-strap through the loose ring on the curb-chain and buckle it on the near side. It need not be tight.
- Check the martingale fitting (page 238). A running martingale should be on the curb/bit rein.

Method 2 *(Figure 82)*

- Put your right hand under the horse's jaw and up round the other side to the centre of his face, just above the nostrils.
- Take both cheek-pieces in this hand, keeping it close to the horse's face.
- Your left thumb opens his mouth and guides his bit in as in Method 1, while your right hand eases the bridle up.
- Using both hands, place the head-piece over the horse's ears.
- Thereafter continue as in Method 1.

Method 2 gives you more control as you are closer to the horse and have your right hand to steady his head and stop him moving about. It is also safer when introducing the bridle to young horses.

Figures 84 and *85* show a snaffle bridle and double bridle correctly fitted.

Finally

- Check the girth and buckle guards adjusting them if necessary. It is a sensible precaution—though not essential—to stand in front of the horse and with your hand behind his knee lift each foreleg in turn and pull it forward, thus smoothing out any wrinkles that there may be in the skin under the girth.
- If you have to leave the horse saddled and bridled, loop the reins under the stirrups and replace the headcollar or halter over the bridle and tie him up, otherwise he may get his reins over his head and tread on them or, worse still, roll and damage his saddle.

84 *A snaffle bridle and drop noseband correctly fitted.*

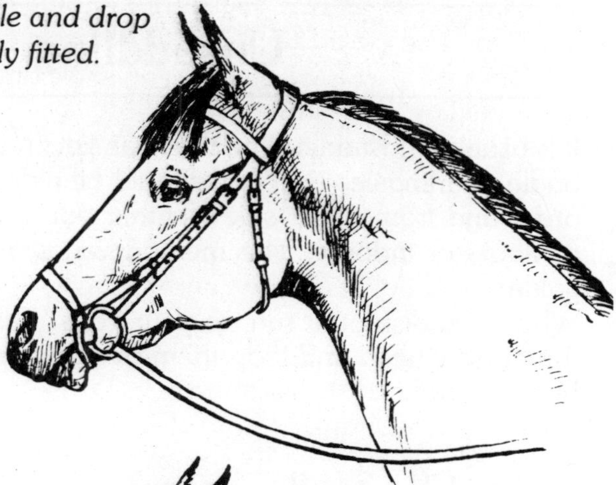

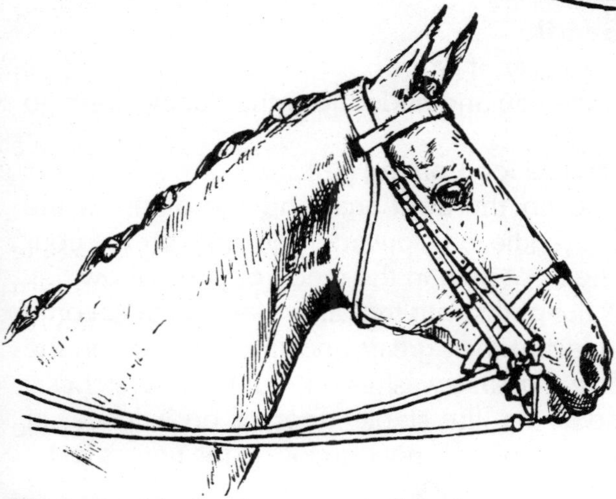

85 *A double bridle correctly fitted.*

- Pick out the horse's feet if he is in a box or stable.
- Complete your own preparations—hat, stick, gloves, etc.
- When ready to move off, untie the horse and remove the headcollar.
- When leading out, put the reins over the horse's head and make sure that the stirrup irons are run up to avoid them getting caught on any projections, particularly in the doorway. Hang the headcollar on a hook or the latch of the door so that it is ready to hand on your return.
- Before mounting, look round once more to be sure that your saddlery is correctly fitted.

Unsaddling

It is usual to unsaddle from the near side in the order: saddle, bridle, martingale—but you should be able to do this in any order and from either side whether you are unsaddling out of doors or indoors. The method varies; the following is a guide:

When you dismount run up your irons. Take the reins over the horse's head and loop them over the arm nearest to the horse's head.

To Take Off a Saddle

- Take off the crupper, if worn.
- Raise the saddle-flap and undo the girths, letting them go gently.
- Slip the martingale loop off the girth.
- With one hand on the front arch and the other on the cantle, slide the saddle off towards you and on to your forearm, with the front arch in the crook of your elbow.
- Take hold of the girths with the other hand as they come over the back, placing their greasy, not muddy, side on the seat of the saddle—grease washes off, but mud scratches.
- Place the saddle on the stable door or on a fence, or lay it carefully on the ground—not too close to the horse's feet.

To Take Off a Bridle

- After a ride horses usually like to rub the sides of their mouth against either their legs or some projection such as door or manger, so the quicker you take off the bridle, the less likely it is to be damaged.
- Have a halter or headcollar near by or hanging over your arm or shoulder, ready to put on.
- Replace the reins over the horse's neck—you had them over your arm while taking off the saddle—so that you can control the horse if he moves.
- Unhook the curb-chain on the near side, leaving the lip-strap done up, to prevent the curb-chain falling on the

ground and perhaps getting lost if it comes unhooked at the other end.

- Undo the noseband and slip off the loop of the standing martingale. If using a running martingale, unbuckle the reins, remove the martingale rings and re-buckle the reins.
- Undo the throat-lash.

You now have everything undone and safe for taking the bridle off the horse's head.

- Place your left hand on the horse's face, well above his nostrils.
- With your right hand, slip the head-piece over the ears and slowly place it on your left forearm, taking care to allow the horse to ease the bit out of his mouth very gently. If it is dropped out quickly he may throw up his head, get caught up with the bit, and hurt his mouth, which he is likely to remember and be difficult about next time. For the same reason, never take off a bridle with the curb-chain or noseband done up.
- Slip off the martingale neck strap and place it together with the head-piece, but not the reins, on your left shoulder, leaving you with both hands free to put on the halter or headcollar.
- Take the reins over the horse's head on to your left shoulder and tie him up.
- Keeping the bridle and martingale over the left shoulder, pick up the saddle and girth, and your stick and carry them correctly (see page 211) to a safe place. In this way you will leave nothing behind and avoid trailing the girth or reins on the ground.
- Put the saddle on a peg or saddle-horse, and the bridle and martingale on the cleaning bracket or other hook.
- Return to your horse and run your hands over his back and girth groove to feel for lumps or soreness.
- Pat the back briskly, but not heavily, to dry it and restore the circulation—not forgetting to deal with both the near and off sides.

The Rugged Up Horse

When you have finished unsaddling, put on a rug and a roller and do up the front strap.

- Look at the corners of his mouth and chin-groove to see if the bit, curb-chain or noseband have rubbed.
- Check that you have left nothing you need in the stable.

Care of Saddlery

Wise and knowledgeable horsemen take great care of their saddlery which they know plays such an important part in horsemanship.

SENSIBLE PRECAUTIONS

Never drop a saddle or allow it to fall to the ground. Not only will the leather be damaged but the tree may be broken, in which case you may not be able to use the saddle again. When not in use, saddlery should be 'put up' (see page 212) or stowed in a safe place. It is easily damaged if left lying about.

When you leave the horse in his box or stall with saddlery on always tie him up. If you do not, he may:

- Damage the saddle by lying down and rolling;
- Rub his bit or bridle against the door or manger;
- Get his reins over his head and tread in them.

INSPECTION, PRESERVATION AND REPAIR

Inspect all saddlery frequently and attend to minor defects. In particular check:

Stitching This deteriorates more quickly than leather, so it is important to check it frequently especially on reins, girths, and stirrup leathers. At the first sign of wear, or if in doubt, take these to the saddler for re-stitching.

Stirrup leathers These should occasionally be shortened from the buckle end to bring wear into fresh holes.

Bits Look for signs of roughness or wear.

The saddle tree Examine this after a fall, or if you have reason to suspect that it might be broken (see page 209).

Leather This becomes dry and cracks unless kept pliable with oil or saddle soap to replace its natural oil. 'Dubbin' and vegetable or animal oils such as castor, neatsfoot or glycerine are good for leather. Mineral oils such as those used for cars and bicycles, and linseed oil are not satisfactory as they become very hard. It is best to use one of the recognised saddle soaps or preservatives sold in saddlers' shops.

When in regular use, saddlery should need only saddle soap and the occasional application of neatsfoot oil to keep it soft and pliable.

When stored for any length of time, it should be protected against damp or dryness with neatsfoot oil or one of the heavier preservatives.

SUMMARY

Daily care Clean and soap all saddlery after use (see below).

Weekly care Completely dismantle it; clean and saddle soap thoroughly, particularly at the bends. Check stitching, stirrup leathers, etc.

Yearly care It is good practice to have all your saddlery looked over by a saddler who will test and put right as necessary, the stitching, the tree and the stuffing.

Storing Completely dismantle the saddlery, dress it with neatsfoot oil or a heavy preservative and wrap it in newspaper which, unlike plastic, allows the necessary circulation of air. The preservative should be renewed periodically, especially if the atmosphere is dry.

Cleaning Saddlery

EQUIPMENT NEEDED

- An apron or overall to protect your clothes.
- A sponge or piece of rough towelling for washing and a chamois leather for drying.
- A small sponge or piece of foam rubber for applying saddle soap.
- A tin of plain, or a bar of glycerine, saddle soap.
- A tin of neatsfoot or proprietary brand of saddle oil.
- A tin of metal polish or impregnated wool.
- Two stable rubbers—one to cover the clean saddle and one for drying metalwork.
- A dandy brush for brushing sweat and mud from serge linings, webbing girths, etc.

- A nail or a sharpened piece of wood to clean curb-hooks and lip-strap 'D's.
- A duster for 'rubbing up' after using metal polish—alternatively a polish-impregnated cloth.
- A rubber or plastic bucket (to avoid scratching metal), filled with cold or tepid water.
- Hooks and brackets on which to hang bridle, girths, leathers, etc; and a saddle-horse *(figure 72)*.

THE CORRECT METHOD

Leather

- Clean thoroughly, using a sponge dampened with cold or tepid water to remove all dirt and grease. Do not soak leather.
- Remove any surplus moisture with a chamois leather, or allow the leather to dry off naturally.
- Apply saddle soap liberally, using a circular movement with a sponge kept for the purpose and used as dry as possible. Concentrate especially on the under, inner or 'rough' side of the leather as this is the most absorbent.
- If using bar-type soap, dip the end of the bar in the water and rub it on the sponge; or, if using soap from a tin, put the damp sponge into the tin. If the sponge lathers this means that it is too wet and will dilute the soap thus lessening its effectiveness.
- Use neatsfoot oil or some proprietary brand of preservative from time to time, if and when you feel that your leather is getting hard and dry and is losing its suppleness. This may happen when it has dried out after a thorough soaking out of doors, if it has not been used for some time, or if it has been kept in too dry an atmosphere. Apply the oil, preferably to the under-side only, with a sponge or small paint-brush kept for this. When it has been absorbed, apply saddle soap.

Leather MUST NOT be washed with soda, hot water or detergent. Nor should it be placed near a fire or hot radiator or in a warm airing cupboard as these will dry it out too quickly, removing the natural and applied oils.

Other Materials

The correct way to clean other materials used in saddlery, such as metal, webbing, nylon, etc, is given in the following paragraphs under the articles of saddlery which include them.

THE PARTS TO BE CLEANED

The Saddle

Certain parts of modern saddles should not be soaped. Your saddler will advise you.

- To hold it firm, place the saddle on the saddle-horse. *Figure 72* shows the ideal shape. It is bad practice to try to clean or keep a saddle on the back of a chair or on any angular object which allows it to rock or rest on the gullet.
- Strip the saddle—i.e. remove girths and girth-guards, stirrup leathers and irons.
- Clean the lining which may be of leather, linen or serge:

Leather Hold the saddle, pommel down, over a bucket so as to prevent water running under the lining or saturating the felt or stuffing. Sponge the lining with cold or tepid water. Dry it with a cloth, sponge or chamois leather. Go over it thoroughly with saddle-soap or occasionally apply neatsfoot oil as described on page 256.

Linen Sponge off or scrub if necessary, keeping the lining as dry as possible.

Serge Brush well with a dandy brush. If it is very dirty, it may be necessary to scrub it, but if you do so it may take several days to dry properly.

- Stand the saddle up to dry—not too near heat.
- Replace the saddle on the saddle-horse.
- Thoroughly clean with cold or tepid water all leather work—the seat, the outside and the underneath of the flaps. Beware of getting it too wet. Remove all accumulations of black grease and dirt—known as 'jockeys'. If this is difficult, carefully use your finger-nail, or hairs from your horse's tail, laid out flat and tied in a knot. On no account use a sharp instrument for this.
- Soap liberally all accessible surfaces, particularly the under-sides of the flaps and sweat-flaps.
- When the saddle is hard and dry, the saddle-flaps should be softened by applying neatsfoot or a proprietary brand of oil to the under-sides from time to time. When this has been absorbed apply saddle soap.

- Do not use oil on the seat or the outside of the flaps as it will stain the rider's breeches.
- *Never use soap on top of dirt.*
- Clean all metalwork with metal polish, or impregnated wool, or a duster.

The Stirrup Irons

- Remove them from the leathers and wash and dry them thoroughly.
- Remove the rubber treads from time to time.
- Clean all metalwork with metal polish, impregnated wool, or a duster.

Stirrup Leathers, Leather Girths and Buckle-Guards

- Hang them up, preferably on a cleaning bracket away from the wall *(figure 72b & c*, page 214) or on hooks on the wall.
- Treat them in the same way as the saddle; wash them clean and thoroughly soap. If the sponge is too wet, lather will fill the holes of the leathers which, unless kept clear with a match or a nail, will accumulate dirt.
- Keep a piece of flannel or blanket which you have soaked in neatsfoot or a proprietary brand of oil between the folds of a three-fold girth.

Nylon or Webbing Girths

- Brush daily or wash with soap, as necessary. Be sure to rinse them thoroughly.

Finally, 'put up' the saddle either with the girth, stirrup leathers and irons hanging alongside it *(figure 71)* or with the stirrup leathers and irons replaced on the saddle and 'run up', and with the girth laid over the seat *(figure 72).*

Numnahs

There are many different types.

- Keep them clean, supple and well-aired.
- Wash, brush, or follow the maker's instructions.

Wither Pads

As for numnahs.

Pad Saddle

- Brush the felt with a dandy brush.
- Wash it, if necessary with soap, to remove grease, and allow it to dry in a warm atmosphere.
- Keep free of hard 'knots' of felt.
- Beware of moth damage and worn stitching.
- Clean the metal and leather-work as for other items of saddlery.

Bridle

Normal clean, without taking it to pieces:

- Hang the bridle up on a cleaning bracket or on a hook *(figure 72)*.
- Undo the buckle at the back of the noseband.
- Let out the cheek-pieces to the lowest hole, noting the holes that you have been using.
- Take the bridle off the bracket or hook and wash the bit with a sponge or piece of towelling in a rubber or plastic bucket of cold or tepid water, taking care not to put the leather parts into the water.
- Check the bit for roughness.
- Wash the top of the head-piece and replace the bridle on the bracket or hook.
- Wash the rest of the bridle, keeping it taut with one hand and washing with the other.
- When dealing with the reins, step backwards and away from the hook and wash down towards the buckle. Then hang them over another hook to keep them from trailing on the floor.
- Wash the noseband.
- Wipe off the surplus moisture with a chamois leather or allow the leather-work to dry naturally.
- Polish the bit (except for the mouthpiece) and the buckles with metal polish or impregnated wool. Do this before soaping or you will almost certainly leave white polish marks on the leather.
- Soap all leather-work thoroughly, putting as much on the under-side as on the top – this is most easily done by wrapping the sponge round the various parts and rubbing it up and down.

To clean when completely dismantling:

This should be done at least once a week:

- Undo all buckles and mounts; hang up the various parts.
- Clean and soap each part separately, taking special care to soap thoroughly the insides of the bends and folds. When soaping there is no reason why, if you prefer it, you should not lay each part on a flat surface and rub the soap in one side at a time.
- If the mounts are hooked billets, undo them by pushing with the pad at the base of your thumb. Once off the billet, the strap is easily pulled out of the keepers. Always push any difficult buckle fastening back through the buckle from above. It is easier to do this than to pull the end.

When oiling If the bridle needs to be oiled, it is best to take it to pieces. Oil each part well. Put it up with the straps in the runners only. When the oil has been absorbed, soap the bridle.

Breast-Plate and Breast-Girth *(figure 81)*

As for other webbing and leather.

Clean the buckles taking care to remove traces of metal polish from the webbing or leather when you have finished.

Martingales, Neck Strap and all other Leather Accessories

As for other leather equipment.

Clean the buckles as for the breast-plate (above).

3
VETERINARY NOTES

The Healthy Horse
The Medicine Cupboard • General Nursing Care
Some External Treatments
Wounds: Types and Treatment • Lameness
Some Common Ailments • Diseases of the Wind
Skin Diseases • Poor Condition

The Healthy Horse

THE SIGNS OF GOOD HEALTH

It is important to recognise the signs of good health and to watch for any deviation from them. You will quickly come to know whether or not your horse is feeling himself.

The following are indications of good health:

- An alert look, with ears pricking to and fro.
- Coat glossy and lying flat.
- Skin loose and supple and easily moved over the underlying bones.
- At rest, no visible signs of sweating except in very hot weather.
- Eyes wide open and bright, with the membranes under the lids and linings of the nostrils salmon pink in colour.
- Eating up well and chewing normally.
- Body well filled out but not gross.
- Limbs free from swellings or heat (i.e. smooth and cool to the touch).
- Standing evenly on all four feet. Resting a hind leg (but not a fore) is quite normal.
- Sound and taking strides of equal length with the weight evenly spread on all four legs.
- Urine fairly thick and either colourless or pale yellow, and passed several times a day.
- Droppings, which will vary in colour with the diet, passed

The Medicine Cupboard

In the well-organised stable, a place should be set aside for the storage of first-aid drugs and dressings. They should be kept clean, and out of the way of small children. Storing bandages, gamgee and dressings separately in polythene bags keeps them clean, tidy and ready-to-hand.

LIST OF SUGGESTED CONTENTS

Animalintex impregnated poultice dressing	
Antibiotic powder	Container or 'puffer' dispenser from the Sulphonamide group
Antiseptic	Cream: 1 jar or tube Saline solution
Bandages	Crêpe and cling-type
Cotton wool	1 packet
Foam rubber or gamgee	1 wide roll
Gauze	1 small packet
Kaolin paste for poulticing	500g (1lb) tin
Safety pins	
Salts	Container of Epsom type
Scissors	Blunt-tipped
Stockholm tar	
Surgical tape	1 roll
Thermometer	Veterinary clinical type
Vapour rub	1 jar
Vaseline or zinc ointment	1 jar
Witch-hazel for galls	1 bottle

Travelling First-Aid Box

- Animalintex poultice dressing
- Antiseptic
- Bandages
- Cotton wool
- Gamgee
- Safety pins
- Scissors
- Surgical tape
- Wound dressing

approximately 8 times daily in the form of damp balls that break on hitting the ground. Their smell should not be offensive. When the horse is at grass, they may be rather looser but should not be as sloppy as a cow's.

- Breathing when at rest at a rate of 8–12 inhalations per minute. This is most easily measured by watching the flanks from behind.
- Temperature: 38°C (100.5°F).
- Pulse 36–42 beats to the minute.

SOME ESSENTIAL PRECAUTIONS

- Visit the horse every day and learn all you can about him. Look for indications of health.
- Feed him regularly and according to the work he is doing.
- Give him regular worm doses (see page 303).
- Have him inoculated against tetanus (lockjaw) and influenza, as advised by your vet (see pages 293 and 296).
- Have his teeth checked regularly as they may be sharp and needs rasping (see page 279). Your vet will do this.
- Pick out his feet daily. Check the condition of his shoes and look for any signs of thrush. Remember: 'no foot, no horse'.
- If the horse is not thriving, seek expert advice having first satisfied yourself that he is not over-worked or under-fed and that his pulse, respiration and temperature are normal. Keep on worrying until you see an improvement.

Taking a Horse's Temperature

Until you have had practice, do not attempt this without the assistance of an experienced person.

- Read the thermometer and, if necessary, shake it down until the mercury is below 37.4°C (100°F).
- Grease the bulb-end with a little Vaseline to make the insertion into the rectum easier.
- Raise the dock with one hand and insert the bulb and two-thirds of the thermometer for half a minute. Grip the thermometer firmly; if you do not do so it can be drawn into the rectum.
- Withdraw the thermometer, release the dock, and read the temperature. Normal temperature is 38°C (100.5°F), but this can vary slightly with different horses.

- Wash the thermometer in cold or tepid water and dip it in disinfectant.

A rise in temperature to 39°C (102°F) indicates that something is amiss, and you should consult your vet.

Taking a Horse's Pulse

The normal pulse rate at rest is 36—42 beats per minute. Young stock and ponies tend to be a bit faster—up to 45 beats per minute.

- Feel the pulse with your fingers, not with your thumb which itself has an artery which you may confuse with the horse's pulse.
- Feel the pulse in one of three places: under the top of the lower jaw, gently pressing the facial artery against the inner surface of the bone (the easiest place); on the artery on the horse's cheek just above and behind the eye; on the inside of the foreleg, level with the knee-joint where the artery crosses immediately over the bone.

Disinfecting Stables and Equipment

When stables and equipment have been used by an infected horse to disinfect them thoroughly is well worth the trouble and expense.

- Remove and burn all bedding, hay and salt.
- Scrub the walls, door, manger, etc, with a strong solution of disinfectant.
- Limewash the walls if possible, and creosote unvarnished woodwork.
- Disinfect all stable utensils, grooming kit, haynets and buckets, in a strong solution of disinfectant according to the maker's instructions. The Cresol type are suitable for stable floors, drains and utensils but, as they are poisonous, mangers and buckets must be thoroughly rinsed afterwards.
- Wash carefully in disinfectant any tack which has come into contact with infection. Then apply saddle soap or oil thoroughly.
- Wash and disinfect all rugs and blankets.

Isolation

To prevent the spread of disease, see page 266.

When to Call the Vet

In the case of severe injuries call the vet immediately.

If in any doubt as to the horse's state of health you should ask the vet to examine him. An early visit may well prevent serious trouble developing.

Do not telephone the vet outside his surgery hours except in an emergency.

Before calling the vet check the following:

General health Take the horse's temperature and study his rate of breathing.

Lameness In the absence of positive signs elsewhere, i.e. heat, swelling, cuts or sores on the legs or cracks in the heels, the cause is most likely to be found in the foot, even if there is no apparent heat there.

If your farrier can find no trouble in the feet, consult the vet. *Meanwhile, on no account work the horse.*

When the vet calls:

- Have the horse ready and make sure that there is hot water and that washing facilities are available.
- Listen carefully to what he says and make a note of his instructions.
- Make sure that you understand his instructions as to treatment, feeding and exercise, and follow them carefully.

General Nursing Care

KEEPING THE HORSE COMFORTABLE

A sick, lame or injured horse should be made as comfortable as circumstances permit, and kept out of reach, but not necessarily out of sight, of stable companions. Horses are creatures of habit—to leave them in their own box without risk of infection to others will be a comfort to them when they are feeling low, as will your own extra kindness and consideration, working quietly to attend to their needs. At the same time, don't fuss them unnecessarily. Above all, keep them warm.

Grooming

Brush the horse over daily to keep him comfortable, but do not allow him to get cold while doing this.

If he has a high temperature or is feverish it is as well to dispense with grooming altogether or to reduce it to a simple wipe over with a damp rubber.

Sponge out his nose, eyes and dock from time to time, which will help to refresh him.

Pick out his feet regularly.

Bedding and Ventilation

The loose box or stable should be kept clean, dry and sweet, with plenty of bedding. If necessary, sprinkle straw bedding with disinfectant powder to prevent the horse eating it, and 'skep out' regularly.

Ventilation must be adequate: plenty of fresh air but no draughts.

Tying Up

If the horse has to be prevented from lying down, put on his headcollar and secure him at eye-level with a shortish rope—no more than 0.6m (2ft) long—which will allow him to move around.

ISOLATION

When efficiently carried out this may prevent the spread of infectious or contagious diseases.

Move the infected horse to a box well away from, but not necessarily out of sight of, the other horses. Near by, keep a supply of forage, stable utensils, buckets, grooming kit and tack for use only on the infected horse.

Ideally, the person looking after the sick horse should have no contact with other horses. If this is not possible, sensible precautions such as keeping an overall or washable stable-coat and rubber boots for wear only when with the sick horse, and frequent washing of hands will lessen the risk of infection.

DIET AND FEEDING

While confined to his box the horse should be on soft food—mildly laxative to keep his bowels working, but not so laxative as to cause diarrhoea.

The fact that a sick horse refuses his feed is always a source of worry to his owner. But nature knows best; the time to encourage him to eat is when he is starting to get a little better and is inclined to feed again. Meanwhile, offer him—at 4-hour intervals during the day—small quantities of tasty, succulent food.

Bran mashes are normally the staple diet for a sick horse, so lay in a supply of bran and *make sure that you know how to mix a bran mash (see page 146).*

See that clean, fresh water and salt in some form are always available. *Never leave uneaten food in the box or stall.*

When the horse starts to work again return him gradually to his normal diet as his work programme builds up.

Suitable foods Hay (meadow hay for preference); bran mashes with some salts, some boiled linseed and perhaps some boiled barley; carrots; apples or other succulent foods; fresh-cut grass providing that it is the right time of year. The grass should be as long as possible and *not* lawn mowings. Only small quantities should be cut at one time as it will over-heat if left in the manger or in a pile on the floor.

Unsuitable foods Hard foods such as oats, bruised barley, beans, high-protein cubes, etc, which are over-heating.

ADMINISTERING MEDICINE

Always follow the instructions supplied with the medicine or seek advice from someone with expert knowledge.

The following are the normal methods of administering medicine:

In the Feed

This is the simplest way if the medicine is palatable and if the horse is not so ill as to be off his feed.

- Dampen the feed, then add the medicine either in liquid or powder form. Mix feed and medicine together thoroughly and offer to the horse at the normal times.
- A wise owner will prepare a food, such as treacle, that his horse particularly likes. Medicines can be more easily disguised when mixed with a favourite food before being added to the normal feed.
- Worm doses in cube form are also available. Most horses will eat these—either in the feed or direct from the hand.

In Drinking Water

This method is only useful if the medicine will mix with the water and if the horse will drink it. It is not a suitable method where automatic drinking-bowls are used.

On the Tongue

This is not as easy nor as effective as one might think. Until you have had practice you should not attempt this without expert assistance.

The procedure is as follows:

- Mix the medicine with treacle or other favourite food to make a stiff paste.
- Hold the horse's tongue out to one side.
- Smear the paste on the back of the tongue or back molar teeth with a smooth, blunt stick or a wooden kitchen spoon.

Cough electuary can be easily and effectively given by one of the following methods:

- By wrapping it in a length of gauze and putting it in the horse's mouth, like a bit, with the ends tied securely to the headcollar and leaving it to be inhaled and absorbed.
- By smearing it along the mouthpiece of the bit and leaving it in the horse's mouth until it dissolves.

If the above methods fail—as with some horses they will—seek veterinary or other expert advice.

LAMENESS

The cause of lameness should be diagnosed (see page 283), if necessary, by your vet and the horse should be exercised only on his instructions.

If the horse is lame in one leg he is likely to rest it, so put a support bandage with gamgee on the opposite leg to help it take the extra weight that it will have to bear.

Provide a good bed to encourage the horse to lie down, but see that it is not so deep as to impede his movements.

A lame horse should not be worked. Remember that rest above all else is essential to recovery.

CARE OF WOUNDS

Carry out the treatment prescribed by your vet conscientiously or, if you have not consulted him, follow the instructions given in the section on Wounds: Types and Treatment, page 274.

Cleanliness is all-important. However, do not interfere unnecessarily with a wound or its dressing as this will slow up the healing process.

Wash your hands before and after dressing a wound. Keep dressings as clean as possible and burn them as soon as they become soiled.

Tearing at bandages and dressings If the horse acquires this habit, seek expert advice on fitting a cradle.

ILLNESSES

If the horse has a high temperature or is feverish, good nursing can do much to help.

Be sure to keep him warm. If an extra rug is needed it should be a light, warm one. Do not over-burden the horse with unnecessary weight.

If the horse is shivering or sweating, a sweat-rug next to his skin under his other rugs will help the air to circulate.

Stable bandages will add to his warmth. They must be removed and replaced at least once a day.

Rubbing his legs and fetlocks with the palms of your hands when the bandages are off will help to restore the circulation.

Ear-pulling, that is to say, grasping the ears at the poll and allowing them to slip gently through your closed hand is comforting to him.

Change his water several times a day. An ample fresh supply is most important at this time; horses do not like to drink warm water.

Good ventilation is vital. There is no illness for which fresh air is not salutary—but draughts are dangerous.

Coughs, Colds and Other Illnesses

These require additional treatment as described in Some Common Ailments, and Diseases of the Wind (see pages 292 and 297).

Skin Diseases

Although some skin diseases do not make a horse really ill, they necessitate isolation, and a special diet may be recommended. It may well be possible to exercise the horse normally. Types and treatment are described on pages 299–300.

Eye Injuries and Infections
Consult your vet and follow his instructions carefully.

EXERCISE

In cases of severe illness follow your vet's advice. In minor cases, use your own good sense.

When a horse has been confined to his box for some time, exercise must be built up gradually. Start with, say, 10 minutes in hand on the first day and increase the exercise period by the same amount each day either in hand or riding him out.

Turning Out
The danger here is that the horse will immediately set off round the field at full gallop—the last thing you want if he has had leg trouble.

If in doubt as to how to deal with this problem seek experienced advice. The following are sensible guidelines:

- Turn him out when he is hungry and wants to eat grass, preferably with a quiet companion and in a small, secluded area on a calm day.
- The first time that he leaves the stable after illness, do not keep him out for more than an hour.

THE GRASS-KEPT HORSE

Wounds may be treated in the field, but if they are open wounds, precautions must be taken to protect them from flies.

Mild lameness may also be treated in the field unless exercise is forbidden.

If bandaging or poulticing are needed they must be carried out in the stable.

If the horse develops a temperature he should be brought in and stabled for treatment.

While he is unwell be sure to visit him several times a day.

Some External Treatments

PRELIMINARIES

Before starting, have all that you need for the treatment near to you in a clean, safe place, where the horse cannot reach it:

- Wash your hands thoroughly.

- Do not move the horse to another box except to hose him, because when ill or lame he is better undisturbed.
- Make sure you have a good light in the box.
- If hosing, position yourself and the horse near to a drain.
- When possible, have an assistant to hold the horse, or secure him by a string loop (see page 103) to something robust—never a drain-pipe or fixture that can be pulled away from the wall.

When you have finished:

- Burn any used dressings.
- Wash your equipment in an antiseptic solution.
- When it is dry, put it all away.

BATHING WOUNDS

This should be done very gently with cotton wool and one of the following:

- Warm water with a few drops of antiseptic added and a proprietary brand of soap.
- A saline solution: 1 teaspoon of salt to every 0.6 litre (1 pint) of warm water. The salt helps to stimulate the healing process, but can cause irritation to the skin.
- 1 part hydrogen peroxide to 10 parts warm water.

Ideally the water should be boiled first and allowed to cool. *Do not use*: sponges for bathing; disinfectants, detergents, or strong solutions of antiseptic which can kill living cells.

For treatment of specific wounds see page 274.

BLISTERING

This is an artificial means of inducing severe inflammation by blistering the skin. It is helpful in cases of strain because the inflammation causes an increase in the supply of blood to the blistered area, which hastens natural healing of the damaged tissues: the pain caused also makes the animal less inclined to attempt movement. *Blisters, normally supplied in paste form, must only be applied on the instruction of your vet who will explain in each case exactly what has to be done.*

Blisters vary in strength from quite mild to severe: the former are 'working' blisters, used in cases of mild strain and as a method of toning up the horse's legs; the latter are used in cases of tendon strain, etc.

If a more severe blister than the 'working' type is needed you should not expect to work your horse for at least three, preferably six, months.

COLD HOSING

The use of a gentle and continuous stream of cold water is a simple and valuable form of treatment for a number of conditions such as the initial cleaning up of some wounds, and the reducing of pain or swelling from strains, etc.

The procedure is as follows:

- Grease the heel thoroughly to avoid soreness.
- Start with a slow, steady trickle of water using it on the hoof first.
- Work gradually up the leg until you reach the injured part.
- The best results will be obtained if hosing is continued for 20 minutes each time and is carried out at least twice a day.
- After you have finished, dry the surplus moisture from the heel and rub in a little more Vaseline or cream to guard against cracks.

HOT FOMENTATION

This may be used after cold hosing and is helpful for ailments where pain or swelling occur, and where a poultice cannot be applied. It is of little value unless continued for at least 20 minutes. In severe cases it may be repeated two or three times a day.

The procedure is as follows:

- Half-fill a bucket with hot water (no hotter than is comfortable for your hand).
- Add a handful of Epsom or common salts to increase the drawing properties.
- Take two pieces of clean blanket, towelling or other thick cloth approximately 60cm (2ft) by 75cm (2ft 6in) and immerse them in the water.
- Take out one piece, folding it either double or in four; squeeze out some of the water and cover the injured part with the cloth until some of the heat is lost.
- Replace the cloth in the bucket, and, taking out the other piece, repeat the process.
- Keep the water hot by adding to it from a kettle.

HOT TUBBING

This works on the same principle as hot fomentation, but with the affected part itself put into the water. It is useful for injuries to the lower part of a limb and to a foot.

The procedure is as follows:

- Pick out the foot and scrub it clean.
- Half-fill a strong plastic or rubber bucket with warm water—no hotter than your hand can bear—and add a handful of Epsom or common salts.
- Place the foot or limb in the bucket and splash water over it with your hand, adding more warm water from time to time.
- Keep the foot or limb immersed for 20 minutes and, in severe cases, repeat the process twice daily.

POULTICING

There are a number of poultices on the market. They can be applied to whichever parts of the body it is possible to attach them.

Their uses are:

- To soothe bruising.
- To reduce inflammation.
- To clean wounds.
- To draw off pus as it forms.

Test all hot poultices with your bare elbow or on the back of your hand. If they are too hot for you to bear they are too hot to apply. Proprietary brands of poultice are normally left on for at least 24 hours and, except where there is poisoning, they will continue to be effective for two or three days.

Some Common Types of Poultice

Kaolin This is most suitable for inflammation and bruising.

The procedure is as follows:

- Heat it up by placing the whole tin—or better still only part of its contents in another container—in a pan of water, leaving the top of the tin open or loosely covered. It is very dangerous to heat the tin with the lid tightly fitted as, when you open it, the contents may well explode in your face and give you a severe burn.
- Spread the kaolin with a blunt knife on to a piece of cloth,

brown paper or tin-foil. In cases of sprains or bruising it can be applied directly to the injured area.

- Alternatively, to heat the kaolin spread it on to a piece of cloth, brown paper or tin-foil, then fold it and put it in a warm, not hot, oven taking care not to over-heat or dry it.
- Once the kaolin has been applied to the affected area, cover it with waterproof material over gamgee or cotton wool, and hold it in place with a crêpe or ordinary bandage.

Kaolin may also be applied cold, in the same manner

Animalintex This has very good drawing properties. It is a prepared poultice which should be laid on a tray or other flat surface. Enough boiling water should be poured over it to ensure that it becomes saturated. Thereafter apply the poultice as for kaolin.

Bran and Epsom salts poultice This is most useful for injuries to the foot and is applied after the shoe has been removed.

The procedure is as follows:

- Bandage the leg. Before applying the poultice smear Vaseline over the heel.
- Heat up a small bran mash, adding a handful of Epsom salts.
- Put enough mash in a white plastic bag of a size that will contain a 6cm (2in) layer of mash.
- Lift the foot, put it in the bag and pack the bran around and under it.
- Put the foot down and fit a poultice-boot. If a boot is not available, use a sack placing the limb in one corner of it and securing the sack by threading a string through it.
- Bandage over the sack to keep it in place.
- If the horse shows a tendency to eat the poultice, seek expert advice.

Wounds: Types and Treatment

There are five main types of wound:

Clean-cut Caused by a sharp instrument, e.g. a knife or a piece of glass.

Torn wounds A tear in the skin or flesh, caused by, for example, barbed wire or a projecting nail.
Puncture wounds Caused by thorns, stakes, nails, etc. The surface puncture is small but the penetration may be deep. They are always serious and particularly so if in the vicinity of a joint.
Bruised wounds Caused by kicks, blows, falls and overreaching.
Galls Caused by loose, badly fitting or dirty tack.

STAGES OF TREATMENT

There are four principal stages in the treatment of wounds:

Stage 1: Arresting the Bleeding

Bleeding is always worrying to the owner, but, unless excessive, is rarely a danger to the horse. If small vessels only are involved, bleeding generally stops of its own accord within 20 minutes.

Excessive bleeding, such as the spurting of bright red blood from a severed artery calls for control, which you should attempt either by bandaging the wound itself or, better still, by applying pressure on the blood vessel above the wound. Your tie or stock can be used as a bandage. To apply pressure to the blood vessel, some small, hard object such as a pebble can be placed inside a handkerchief and tied on to the leg as an emergency measure while out riding. It is dangerous to leave this type of bandaging on for more than 20 minutes.

Stage 2: Cleaning Up

This is by far the most important part of any treatment. The time and energy expended in the preliminary cleansing of a wound is of much greater value than any cleansing done later on. Many wounds heal quickly and well if the preliminary cleansing and removal of foreign matter has been carried out with thoroughness. First, clip the hair in the vicinity of the wound if this is possible. Then decide how you are going to clean it—by washing and/or by poulticing.

Washing There are two principal ways:

Bathing (see page 271).

Cold hosing (see page 272).

Poulticing If the wound is a puncture type or you suspect that some dirt may still be in it you should apply a poultice.

Stage 3: Dressing

If small, the wound should be liberally dusted with an antiseptic or antibiotic powder or, in some cases with a healing cream.

If extensive, cover it with an impregnated dressing or bandage over gamgee.

In the case of bruised wounds, the dressing may be a poultice applied after cold hosing has been carried out to reduce swelling and soreness.

Stage 4: Protecting (Bandaging)

This is not always necessary, and in any case is generally only possible on the lower parts of the limbs.

Cover the wound with lint and gamgee, and bandage lightly to allow room for swelling.

Subsequent Dressings

Repeat Stages 3 and 4 daily. If the wound is still dirty repeat Stage 2 also.

In some cases, healing is quicker if the wound is not disturbed and the dressing left on for several days, but you should take expert advice before doing this.

A wound that is becoming septic will produce the following symptoms:

- An increase in lameness.
- Swelling and heat, particularly swelling that creeps up the limb.
- Obvious discomfort and possibly sweating and loss of appetite.

In all such cases you should seek expert advice without delay.

Additional Measures

These are described under General Nursing Care, page 265. Note particularly:

Diet This should be adjusted if the horse's work is to be reduced.

Anti-tetanus injections Arrange for these to be given, especially in the case of puncture wounds.

Antibiotics These kill or inhibit the multiplication of germs within a wound so effectively that disinfectants have been largely superseded. Antibiotics should always be administered in cases of puncture wounds. Recovery-time is reduced and the more severe after-effects of the injury avoided. Antibiotic drugs can only be obtained through a vet.

COMMON WOUNDS AND THEIR TREATMENT

Clean-Cut and Torn Wounds

Follow Stages 1 to 4 (pages 275-276).

Stitching This can greatly help the healing process if carried out within a few hours of the injury, but it requires the attention of a vet.

Puncture Wounds

These are a dangerous type of wound and are generally troublesome, penetrating deeply and leaving small openings. Festering may occur after a day or two and, as the pus may well have difficulty in escaping, the wound may be very painful.

Treatment

- Apply a poultice or, if this cannot be attached, a hot fomentation (see page 272).
- A bran poultice or hot tubbing are most suitable for the feet.
- Puncture wounds must heal from the deepest part so make sure that the entrance does not close up too soon.
- Puncture wounds to the foot can be washed out with a disinfectant providing that the instructions are carefully followed.

Bruised Wounds

These can be very painful and they require treatment even if the skin is not broken.

Treatment

- For the first 24 hours administer cold treatment: cold hosing (page 272) followed by a cold kaolin poultice (page 273) to clean, stop internal haemorrhage and reduce inflammation.
- The next day begin warm treatment: warm poultices or, if a poultice cannot be attached, fomentations, for up to 3 days

to absorb any discharge, to soothe the bruise and to reduce inflammation.

- Thereafter treat with zinc or castor oil ointment.

Note. In cases where the injury has taken some time to form, e.g. capped elbows and saddle galls, and when the haemorrhage is likely to have stopped, then start with warm treatment.

Galls

Girth galls Sores and/or swellings on the soft skin behind the elbows caused by the friction of a girth that is too loose, too hard, too broad or has caked mud on it. See also Girth Itch.

Swellings, if unnoticed, may form hard lumps which may also become raw. Ninety-nine per cent of girth galls are the result of bad management. See Saddling Up, page 244.

Treatment

- If the skin is unbroken, harden it with applications of saline solution or witch-hazel lotion.
- If the skin is broken, wash it with a weak saline solution—1 teaspoon salt to 0.65 litre (1 pint) warm water; dry with cold cotton wool swabs; dust with sulphonamide powder.
- *On no account use a saddle again until the gall is healed and the swelling has gone down or is painless.*

Girth itch In some cases the sores are not true galls but are caused by girth itch, a form of ringworm resulting from an infected girth. If in doubt, seek expert advice and treat as for ringworm.

Saddle galls (or sore back) These consist of any injury, from a slight rub to a severe swelling, abrasion or 'sitfast', caused by friction or pressure from the following:

- Badly-fitting saddle or roller that pinches the withers.
- Saddle overdue for re-stuffing.
- Broken saddle-tree.
- Bad riding—i.e. the rider rolling about in the saddle.

Treatment

- Identify the cause and rectify it.
- Reduce inflammation or swelling with hot fomentations.
- Heal open sores as for girth galls.
- Harden as for girth galls.
- On no account use a saddle again until the gall is healed and the swelling has gone down or is painless.

'Sitfast' This forms when a portion of the skin dies because the circulation has been cut off by pressure. The portion in question is like an inverted cone and becomes surrounded by pus. In these circumstances a vet must be called in.

Bit and Mouth Injuries

These are injuries to the bars or roof of the mouth, the tongue, the cheeks, the lips, the corners of the mouth or the chin-groove. They are caused by wrong-sized, badly-fitting or worn bits; rough riding; ragged molar teeth.

Treatment

- Do not use a bit until the injury has healed. If the chin-groove is affected, ride without a curb-chain.
- Wash out the mouth after feeding with a warm saline solution—1 teaspoon salt to 0.65 litre (1 pint) water.
- Have the molar teeth checked and filed if necessary.
- Change or adjust the bit.

Mouth wounds, other than bruising, are generally quick to heal.

Broken Knees

This term is used to describe injuries to the surface of the knees caused by the horse stumbling or falling. The injury varies from a slight abrasion of the skin to actual exposure of the bone.

The extent of the damage may be:

- A slight graze, just removing the hair, which leaves no visible damage.
- Deep-seated damage to the tissue, with the roots of the hair destroyed and probably gravel or dirt in the wound, resulting in a scar and possibly a permanent bump of proud flesh.
- Extensive damage affecting ligaments or even the joint-capsule which could result in loss of joint-oil. There is then a risk of infection which may cause disease to the joint and a permanently stiff knee.

Treatment

- *Graze:* Follow Stages 2 to 4 (pages 275-6), but bear in mind that there may also be bruising to the knee.
- *Extensive damage:* Consult a vet and follow his instructions closely.

An amateur probing about in a wound close to a joint can easily increase the damage.

Bruised Sole (See page 284).

Cracked Heels

These are open cracks in the heels which, if neglected, become deeper and longer, and will make a horse very lame. They are caused either by an infection such as eczema, or, more probably, by chapping of the skin resulting from failure to dry the heels properly after work or washing.

Prevention Sometimes cracked heels are due to neglect. Some horses are more prone to them than others. The following precautions are helpful:

- Clip out their heels regularly.
- Apply a healing ointment before work, and on return from work after first cleaning and drying them.

Treatment

- As for a wound (Stages 2–4), but using a proprietary ointment. Providing that the horse is sound, after applying the ointment there is no reason why you should not work him if you follow the advice for prevention.

Mud Fever

This is an irritation of the skin caused by wet and mud, and is particularly common in winter. The skin of the legs and stomach becomes tender and scaly and, in severe cases, temperature and fever result.

Ninety-nine per cent of cases are due to one's own neglect or mismanagement. Some areas of the country are more prone to it than others. Wherever you live, if you wash your horse or bring him in wet after work and you do not dry him thoroughly, you are asking for trouble.

Prevention

- Bring the horse in from work as dry as the weather allows.
- Dry out his heels.
- If his legs are wet and/or muddy, bandage them and leave the bandages on until next morning when the mud will brush off easily.
- Remove all mud once the horse is dry—however tired

you or he may be—using straw, your hand and a very soft brush—not a dandy brush.

Treatment

• Consult your vet, and feel thoroughly ashamed.

Pricked Foot

(See page 285).

Under-Run Sole

(See page 287).

Self-Inflicted Wounds

Brushing This is the term used for wounds on the inside of the fetlock joint, and sometimes the coronet area, that are the result of a blow from the edge of the shoe or rough clenches on the opposite foot.

There are many possible causes: bad conformation and faulty action; a raised clench, ill-fitting or heavy shoes; weakness due to youth, old age, over-work or under-feeding.

Prevention

• Trace the cause and eliminate it if you can.

• If you suspect the shoes, consult your farrier who will probably fit a feather-edged or some other suitable shoe (page 165).

• Meanwhile use a brushing boot *(figure 48)*, and continue to do so if in doubt.

Treatment

• As for Bruised Wounds (page 277).

Over-reaches These are wounds on the bulbs and sometimes the heels themselves, caused by the hind foot striking into them. Normally over-reaches occur only when galloping or jumping, and they are invariably accidental, particularly in deep going.

Prevention

• If you suspect either the shoes or faulty action, consult your farrier who may fit a shoe with a 'safed off' toe (page 164). Meanwhile, use over-reach boots *(figure 48)* and continue to do so if you are in any doubt.

Treatment

• As for Bruised Wounds (page 277).

Speedy-cuts These are wounds on the inner and lower side of the knee, caused by the inside of the toe of the opposite fore leg. They are very serious, and veterinary advice must be sought.

Prevention

- Trace and if possible eliminate the causes, which are the same as for Brushing (page 281).
- Meanwhile use a speedicut boot *(figure 48)*.

Treatment

- As for Bruised Wounds (page 277).

A horse that really speedy-cuts is unsafe to ride.

Striking into The same as over-reaches but higher up the leg and probably cutting into the tendon—a very serious injury.

Prevention

- As for over-reaches, but use a tendon boot *(figure 48)*.

Treatment

- As for over-reaches.

Treads Wounds to the coronet region that are the result of a horse being trodden on or of treading on himself, usually while travelling.

Prevention

- When travelling use coronet boots *(figure 48)*, or bandage with gamgee to below the coronet.

Treatment

- As for Bruised Wounds or if studs are involved, Puncture Wounds (page 277).

Lameness

Lameness is common, and although it is for the most part a matter for the expert, every rider should have some understanding of the subject and should certainly be able to recognise when his horse is lame. A thorough knowledge of the horse and his action is most helpful in this respect.

When riding, unevenness of step is fairly obvious, and if you have any misgivings the horse is almost certainly slightly lame.

Dismount at once and look for stones which may be lodged in his foot, and then look him over for any other obvious cause.

If you can find no cause for it and he remains lame, lead him home on soft ground, or telephone for help.

LOCATING THE SOURCE OF LAMENESS

Identifying the Lame Leg

Take the horse on to some hard, smooth ground, walk him up and down, then trot him on a loose rein. If the lameness is severe it will show at the walk.

If he is lame in front he will drop his head as his sound leg comes to the ground, raising his head again to take the weight off the painful leg. If lame on both forelegs he will tend to keep his head raised and to shuffle rather than stride out.

If he is lame behind he will similarly try to take the weight off the painful leg by leaning over to the sound side, causing him to lean more on the sound leg. He may also appear to hitch up his quarters (and hock) or drag his toe on the lame side.

If you suspect lameness in a foreleg and can find no cause for it, it may be that the trouble is in the hind leg diagonally opposite. Watching the horse at rest in the stable sometimes gives a clue to obscure lameness. Although standing square,

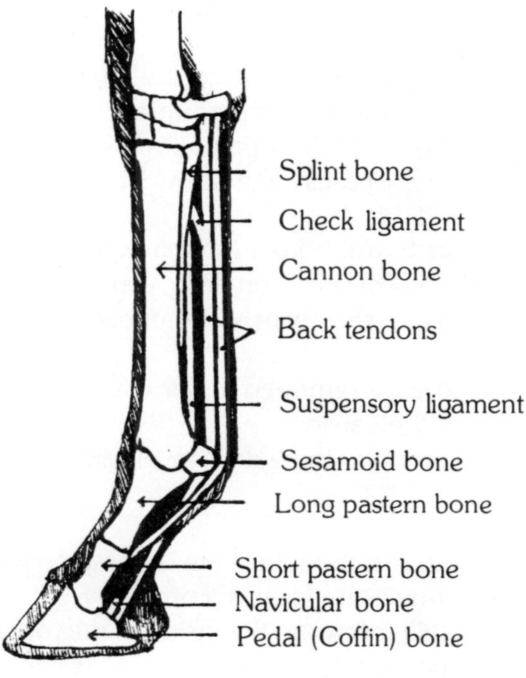

86 *Section of foreleg showing the parts liable to injury.*

the pastern of the lame leg may be slightly straighter than that of the sound one, or he may be 'pointing' the lame foreleg in front of the other.

Finding the Trouble Spot

As most lameness stems from the foot, begin your examination there.

- Check for anything sharp that may be lodged in it; any displacement of the shoe; any sign of heat—do this with the palm of your hand on the wall of the hoof. There will be no swelling as the foot cannot expand.
- Run your hands down all parts of the leg. Any signs of heat, pain or swelling—unless a permanent enlargement—will indicate the seat of the trouble. In obscure cases there may be none of these indications.

THE FOOT

Bruised Sole

This will affect not only the horn of the sole, but also the deeper, sensitive part underneath.

A true bruised sole is usually caused by picking up a stone in the foot, or by hard or rough ground.

The sole may be pink or painful when pressure is applied.

Except in severe cases the horse will be lame only on hard or uneven ground.

Prevention Some horses have naturally very thin soles or flat feet and are prone to bruising, especially on stony tracks. If this is the case, shoe with a leather pad that covers the sole.

Treatment

- Have the shoe removed and the foot examined carefully to ascertain the cause.
- Poultice for up to three days to alleviate the pain.

Corns

These are bruises of the sole in the heel region just beneath the heels of the shoe—the 'seat of corn'.

They generally occur on the inside of the fore feet only. The bruise may be simple or severe with festering. There will normally be heat in the heel region.

The cause may be: badly fitting shoes—too short, too close, or too narrow at the heels; shoes left on too long; excessive paring of the heels and the bars of the hoof which weakens them.

Prevention

- Once a horse has had corns they are always liable to recur.
- Shoe with a surgical shoe—'three-quarter' or 'seated-out' (page 166) to remove the pressure.

Treatment

- Send for, or visit by box or trailer, your farrier who will: remove the shoe; pare out the corn; re-shoe with a surgical shoe (page 166).
- If the farrier is not available, send for your vet.
- Apply a poultice (page 273).

Laminitis

This is a fever in the feet due to intense congestion of the sensitive structures lining the walls of the hoof. It is a serious disease and very painful, as the foot cannot expand to allow for the swelling. Ponies are particularly susceptible to it.

It affects all four feet—never a single foot. The horse will stand on his heels with his fore feet thrust forward, and be very reluctant to move.

It is caused by: too much food and not enough work; too much heating food, particularly when the horse is stabled; too much lush spring grass, especially with fat horses; too much trotting on the roads or fast work on hard ground.

Treatment

- Consult your vet.
- Meanwhile, stand the horse in the stable on bedding that he won't eat, or on a bare floor with water, but *no food.*

Pricked Foot/Nail Binding

This is pain in the sensitive structure of the foot due to a nail being driven too close and causing pressure on it. Only your farrier or vet is likely to spot this.

Treatment

- When the nail is removed the horse will usually go sound unless poisoning has set in. If this is the case treat as a puncture wound (page 277).

Navicular

This is a serious arthritic condition for which, at the moment, there is no known cure. In simple terms the navicular bone becomes rough, making the tendon in contact with it sore.

It is usually caused by concussion, but it can be hereditary.

Symptoms

- Obvious lameness which in the early stages wears off as the horse warms up.
- Heat in the foot.
- 'Pointing' one foreleg—or both alternately—when standing in the stable.

Treatment

- Consult your vet.

Pedal Ostitis and Sesamoiditis

Similar to navicular but the damage is to the pedal or the sesamoid bone.

Treatment

- Consult your vet.

Sand Crack

This is a crack in the wall of the hoof running down from the coronet. It usually occurs on the inside quarter or the toe of the hind hoof and is due to brittle feet, mainly caused by either poor condition or too much rasping of the wall of the hoof.

Treatment

- If the horse is lame, consult your farrier or vet.
- If the horse is not lame, your farrier will isolate the crack by making a groove across it with a hot iron.
- Rub one of the proprietary brands of hoof-dressing into the coronet to stimulate the growth of the horn.

Seedy Toe

This is an infection of the foot usually appearing at the toe and found by the farrier when shoeing the horse.

It is caused by a blow anywhere on the hoof or by faulty shoeing.

Treatment

- Your farrier will advise you.

Stone in the Foot

Treatment

- Remove the stone. If the horse is still lame, treat as for Bruised Sole (page 284).

Thrush

This is a diseased condition of the cleft of the frog. It has a very strong and offensive smell.

It is caused in most cases by standing in ill-drained stables on urine and dung and not having the feet picked out.

Treatment

- Remove the cause.
- Pick out the feet at each 'stables'.
- Scrub the frog with medicated soap and water.
- When dry, dress with antibiotic spray for a few days, followed by an application of Stockholm tar held in place with tow or cotton wool.
- Continue until cured.

Under-Run Sole

This is pain in the sensitive structures due to the foot having been penetrated by a sharp object such as a nail, flint, or piece of tin or glass.

Festering may occur and, as the pus cannot escape nor the foot expand, intense pain is likely.

Treatment

- As for Puncture Wounds (page 277).

SPRAINS

Sprains of the tendons, ligaments and muscles round the joints are common causes of lameness. They may occur under any conditions, but do so usually while jumping or galloping in heavy going and particularly when the horse is tired.

Serious sprains that put a horse out of work for many months seldom occur without warning, which comes in the form of slight heat or swelling.

It is therefore very important to keep a close eye on the horse's legs, feeling them carefully with your fingers and the palm of your hand, particularly before work. The better you know the horse's legs when he is sound and well, the more easily you will be able to detect trouble.

Tendons Tough fibrous cords, slightly elastic, attaching muscle to bone and giving support to joints. When tendons are sprained, there will be pain, swelling and lameness.

Ligaments Strong fibrous bands—like rope—connecting bones to bones. With the exception of the suspensory ligament, they are non-elastic and do not stretch.

A sprained ligament may cause a little swelling. Lameness will follow immediately if a check ligament is affected and after a short period if the sprain is in a suspensory ligament *(figure 86, page 283)*.

Treatment

- If you feel signs of heat or swelling, seek expert advice.
- Meanwhile cold hosing (page 272) for the first 24 hours followed by a poultice (page 273) will help reduce the inflammation.
- Put a support bandage on the other leg.

Remember that a slight sprain can easily become very serious if you continue to work the horse, and that a serious sprain can well put him out of action for six months or a year, and involve blistering.

Curb

This is a thickening of the ligament on the hind leg approximately a hand's breadth below the point of the hock which, when forming, can cause lameness, though many horses have it through life without going lame. It is caused by strain and/or bad conformation of the hock. The area affected appears to be bowed and there will be heat while the curb is forming.

Treatment

- Seek expert advice.

BURSAL ENLARGEMENTS

Bursal enlargements are soft visible swellings caused by:

- Strain—sudden or sustained pulling on a joint or the tendons or ligaments associated with that joint.
- Injuries—blows or abrasions.

Strain

Windgall Swelling just above and to the sides of the fetlock joint. Often painful and a cause of lameness when first formed, but generally giving little trouble later *(figure 87)*.

Bog spavin Soft swelling on the inside of and to the front of the hock *(figure 87)*.

Thoroughpin Soft swelling in front of the point of the hock *(figure 87)* which can sometimes be pushed through from one side of the leg to the other.

Capped knee Soft swelling where the tendon passes over the front of the knee.

Treatment

- Although these are not very serious they are a warning of strain and it is important to try to get rid of them.
- Rest the horse while there is any soreness.
- Cold hose as described on page 272, followed by hand massage with some embrocation.

Injuries

Capped elbow Soft swelling under the skin on the point of the horse's elbow. It is usually caused by either lack of bedding

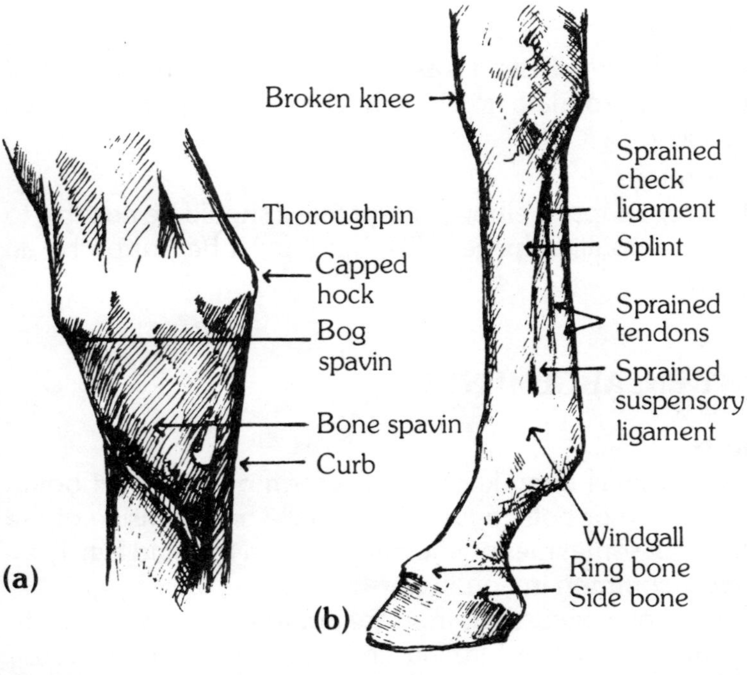

87 *The seats of lameness.* **(a)** *The inner aspect of the hock.* **(b)** *The foreleg.*

or by rubbing with the heel of the shoe on the elbow while lying down.

Prevention

- Remove the cause. If it is the shoe, your farrier should round the heel.
- Meanwhile fit a 'sausage' boot round the fetlock to prevent further contact.

Treatment

- Treat as for a Bruised Wound except that since the injury is likely to have been forming for some time, go straight to hot fomentations (page 272) to soothe and reduce inflammation.
- If the skin is broken, cover it with antiseptic cream between fomentations.

When the injury is no longer painful, massage it with liniment.

Capped hock Soft swelling under the skin on the point of the hock. It is usually caused by lack of bedding; kicking at the sides of the loose box or trailer, etc.

Prevention

- If the habit persists, it is as well to pad the area of the box or where appropriate to fit a hock-boot—especially when travelling.

Treatment

- As for a capped elbow, except that it should be possible to attach a poultice (page 273), but it must be put on by an expert.

BONY ENLARGEMENTS

Splints

These are small bony knobs which form on the splint bone, cannon bone or both *(figure 87)*, usually on the inside of the foreleg, but sometimes on the outside or on the hind leg. They are most common in young horses.

Caused by concussion, they are apt to be very sore and to bring lameness while forming; thereafter they seldom cause trouble unless appearing in the knee, or pressing on the suspensory ligament.

The further forward they are on the leg the less troublesome they will be.

Treatment

- Cold hose three times daily (page 272).
- Bandage with cooling lotion on gamgee or lint to help soothe the soreness, or apply kaolin poultice (page 273).

If sound at the walk, walking exercise may be given, but too much work while the splint is forming may result in a very large splint.

If the soreness persists beyond a week you should seek expert advice.

Bone Spavin

A bony enlargement on the lower part of the inside of the hock *(figure 87)*; caused by strain.

Lameness shows when the horse first moves off but diminishes with exercise. Movement of the hock is restricted by it, and the horse will tend to drag his toe.

Treatment

- Consult your vet.
- Meanwhile hot fomentations followed by cold hosing (page 272) will help to reduce inflammation.
- Blistering may be recommended.

The horse is unlikely to be fit for any serious work for several months.

Ringbone

This is a bony enlargement of the pastern bones *(figure 87)* either 'low' on the coronet or 'high' above the coronet.

Caused by concussion or injury. It is slow to form but heat will be evident.

Treatment

- Consult your vet.

Sidebones

Bony growths on the cartilages of the foot in the area of the heel *(figure 87)*.

Caused by concussion or injury. Heat and lameness will follow.

Treatment

- Consult your vet.

Some Common Ailments

SIMPLE FEVER

A rise in the horse's temperature above the normal.

It is caused by either a germ or nervous reaction to pain.

Always take the horse's temperature if you think he looks 'off colour' or if he shows the following symptoms:

- Refusing his feed.
- Dull and listless.
- 'Staring' coat.
- Quickened breathing or shivering.

Treatment

- Consult your vet.
- Antibiotic drugs will quickly check infections and germs, e.g. pleurisy or pneumonia, before they cause real harm, but you will need a vet's prescription for them.
- Meanwhile, and thereafter, abide by the rules for good nursing (page 265).

COLD IN THE HEAD

Similar to the human type. Shows as a yellow/white discharge from both nostrils.

It is caused by infection or exposure; simple catarrh, which often develops as a result of lack of fresh air either in the stable itself or on a long journey in a stuffy box or trailer; change of conditions when the horse is first brought up from grass.

Treatment

Take the horse's temperature, and if it is above 39°C (102°F) consult your vet (see page 263).

If he has no temperature:

- Stop work and isolate.
- Arrange for plenty of fresh air (not draughts) in the stable.
- Rug up and put on stable bandages.
- Feed off the ground to allow the nostrils to drain.
- Feed mainly bran mashes.
- Keep the sick horse's water bucket away from other horses.
- Clear the nostrils several times a day with cotton wool (*not* a sponge) which should afterwards be burned.

COUGHS

There are three main types:

Those associated with a common cold They are in the form of laryngitis or sore throat.

Virus, caused by infection This can be very tiresome and persistent.

Allergy Some horses are allergic to hay and straw and will cough as soon as you bring them into a stable.

Treatment

- On no account work the horse when he has a cough, except on the advice of your vet. To do so will at best prolong the cough, and at worst cause permanent damage to his wind.
- Take his temperature (page 263).
- Observe carefully the rules for good nursing (page 265).

Common cold type As for a cold in the head but with cough electuary—a thick, treacly proprietary mixture. For administration of medicines see page 267.

Virus Consult your vet. He may also advise inoculation for future prevention.

Allergy Make sure there is good ventilation. Feed laxative diets while converting to hard food. Bed down on peat or shavings. Give the horse some grazing and feed only the minimum of hay, well shaken up and soaked. If the allergy persists consult your vet.

EQUINE INFLUENZA

This is caused by a virus and it is extremely infectious and contagious.

Symptoms

- A rise in temperature up to as much as 41°C (106°F).
- A dry shallow cough.
- Signs of exhaustion and depression.
- A slight discharge from the nose.
- Loss of appetite.
- Inflammation of the eyes, gums and other membranes.

Prevention (Always better than cure.)

- Make sure your horse is immunised.
- Horses in good health and under good stable-management are less prone to the virus.

- If there is an epidemic in your area:

Avoid close contact with other horses as much as possible, particularly when travelling, out hunting, or at a show, etc; never allow your horse to drink from a water trough, or to graze while at a show.

Treatment

- Consult your vet, treating it meanwhile as for a cold and cough.

STRANGLES

This is a contagious disease mainly of young horses, but sometimes of older ones which have escaped it in their youth.

It is always preceded by a sharp rise in temperature, possibly as high as 40.5°C (105°F); then come the following symptoms, which vary in degree with different horses:

- Profuse nasal catarrh.
- Formation of an abscess in the jowl region.
- Signs of great distress, with the horse off its feed and swallowing with difficulty.
- The throat region tense and enlarged.

In due course the abscess between the jaw bones bursts and drains, giving immediate relief.

Treatment

- Consult your vet.
- A peculiarity of the disease is that in the majority of cases the final outcome is better if it is allowed to run its course unchecked by antibiotics.
- Isolate the horse (page 266).
- Attend most carefully to the principles of good nursing (page 265).

BASTARD STRANGLES

More serious but not as common as strangles.

The infection penetrates the blood stream and abscesses form on other parts of the body and internal organisms.

Treatment

- As for strangles.

AZOTURIA/'SET FAST'

A peculiar disease, the exact cause of which is not known.

Symptoms

- The muscles of the hindquarters become tense, hard and painful.
- The horse is distressed and the stride of the hind legs shortens giving the impression of lameness.
- If forced to keep moving the horse is likely to get worse and will probably stagger and fall.

It can happen to any type of horse at any time of year, but most often to one that is fit and hard and has had a rest day or has been confined to his stable on a full working diet.

Prevention

Some horses are more liable to the disease than others, and once they have had an attack it is likely to recur.

- Make sure that you don't feed more concentrates than are needed for the work the horse is doing, and consider adding a small quantity of bicarbonate of soda to his water. Take expert advice on this.
- When he is to have a rest day on a full diet, give him a bran mash with a handful of Epsom or Glauber's salts the night before.
- Always work slowly for at least 20 minutes before going on to faster work.
- With a horse that is liable to azoturia, it is better to turn him out in the field for an hour or so daily, provided that conditions are suitable and that he is accustomed to it, than to give him no exercise at all.

Treatment

As soon as you recognise the symptoms:

- Dismount, loosen the horse's girths and cover his loins with your coat.
- Allow him to rest, if possible in a sheltered place.
- If possible contact a vet before attempting to move him further.
- When the horse is able to move, walk him home if it is near by and if he is willing.
- Failing this, he must be taken home in a trailer or stabled in a nearby stable, shed or barn.

On arrival home consult your vet. Meanwhile:

- Rug the horse up and keep him warm and comfortable.
- Give him plenty of water but no food.

- Encourage him to stale, as this will bring some relief.
- Massage the affected muscles and apply hot packs (e.g. hot water bottles).

COLIC

This is a vague term used when a horse has any form of abdominal pain, and is similar to human stomach-ache.

The most common causes are: too much concentrated food especially when fed too soon after hard work; unsuitable food, e.g. new oats or hay in too large quantities, mouldy hay, poisonous foods, over-heated grass-mowings, etc.; digestive trouble, caused by bolting food or too rapid changes of diet; stoppage in the intestines; twisted gut—a very serious condition which may be fatal.

Symptoms

- General uneasiness accompanied by an increased rate of breathing.
- The horse is off his feed.
- Restlessness, looking round at his flanks; kicking at his belly and getting up and down, and rolling.

Treatment

- Colic may be very serious and if the symptoms don't start to subside after 20 minutes or so you must consult your vet.
- When you first notice them:

Remove food and water from the box.
Bed the horse down with a good deep bed.
Watch him carefully to try to diagnose the cause. On no account leave him unattended until he is settled.

- If he is in danger of getting cast when rolling or if he rolls persistently, it is better to take him out and walk him quietly round, making sure you don't allow him to get chilled.
- If the symptoms subside, allow the horse to rest for an hour or so and then offer him a small bran mash.

TETANUS

This is a very serious disease, caused by a germ which lives in the soil and gets into the body through a wound.

Symptoms (They show about ten days after infection)

- General stiffness.
- High temperature.
- Standing with nose and hind legs thrust out.

- Membranes of the eyes extending over the eyeball.
- The jaws becoming locked in the later stages.

Prevention Though cure is extremely difficult, prevention in the form of inoculation is not. You should have your horse inoculated each time he sustains a wound or, far more satisfactory, have him immunised for life with a series of injections followed regularly by 'boosters'. When you buy a new horse, arrange immediately to have it immunised for life unless this has already been done. Keep a careful record of any such injections.

Your vet will advise you.

Treatment If the horse is to be saved, immediate treatment by your vet is essential, so if you are in any doubt consult him at once.

Diseases of the Wind

Diseases of the wind are of two main types:

- Broken wind.
- Roaring and whistling.

The first is a condition of the lungs, and the second an obstruction of the larynx.

BROKEN WIND

This is a chronic condition of the lungs which may result in collapse of the lung structure. It is caused by: strain due to either over-feeding before fast work, or fast work when the horse is gross and unfit; after-effects of some lung disorder such as pneumonia; allergy due to the presence of fungi in dusty hay or straw.

Symptoms

- Persistent 'graveyard' cough—deep and hollow and appearing to come from the belly region.
- Heaving of the flanks and double movement when breathing out.
- Distress when moving.
- Inability to perform fast work.

Treatment

- Seek expert advice.

- Given time and rest, there may be a considerable measure of repair to the lung tissues—the cough becoming less marked and less frequent.
- Special feeding and carefully controlled exercise can help to keep the horse in work. With asthmatics the diet should not contain hay, nor, in particular, dusty foods.
- It may well be best to work the horse from grass.
- Consult your vet about exercise and feeding.

ROARING AND WHISTLING

These are abnormal noises made by a horse when moving fast. They are caused by paralysis of one of the nerves of the larynx restricting the passage of air.

Once established, the condition will not cure itself, and hard fast work will make it worse.

Treatment

There are two alternatives, both involving surgery:

- The Hobday operation to clear the obstruction in the larynx may help but is not often successful.
- Tubing. A silver tube is inserted in the throat below the larynx. The horse then breathes through the tube. This is a noisy and messy operation but it affords considerable relief with no apparent distress to the horse.

HIGH BLOWING

This is a noise made when breathing out, caused by a vibration in the nostril. It is not a disease nor an unsoundness, and should not be confused with whistling.

Well-bred horses are more likely to be 'high blowers'. Those so affected tend to make more noise when they are excited.

TESTING A HORSE'S WIND

The object is to hear if he 'makes a noise' as he takes in his breath. This is best done at the trot and the gallop, preferably when going uphill.

It is probably easier to hear a horse 'whistling' while riding it, but it is also possible to hear it from the ground if the horse just walks past or is brought straight up to you while he is still breathing strongly.

Skin Diseases

LICE

These are sometimes found on horses with long coats at grass, particularly around the month of February. They can also be acquired from other animals. Horses in poor condition are likely to be seriously affected by them.

Types

The blood-sucking species Found at the base of the horse's mane and tail.

The biting species Found on the lower part of the body.

Symptoms

The horse will be itchy, constantly rubbing the affected areas, leaving bare patches of skin. If affected by the blood-sucking species, the horse will lose condition.

Treatment

- Apply louse powder liberally from poll to croup and on other affected areas. It is easier, but not essential, to do this when the horse is clipped out.
- Repeat after ten days to kill off the nits which have subsequently hatched.
- Disinfect thoroughly the stables and equipment used. If the lice are resistant to louse powder, consult the vet.

SWEET ITCH

This is an irritable condition of the skin on the crest, withers and croup region, causing the horse to rub these areas raw. It appears in early summer and generally disappears in cooler weather.

The cause is not fully understood. For many years it was thought that an animal with this condition was allergic to certain spring and summer grasses, but it is now thought to be caused by a midge bite.

Treatment

The condition, which can be confused with lice, should be confirmed by your vet, who will prescribe the necessary treatment. Meanwhile stabling the horse keeps him away from the grass and the midges.

RINGWORM
A highly contagious condition which can also affect humans.

Symptoms

It shows on the skin as circular patches about the size of a 10p piece from which the hair completely pulls away.

Treatment

- If any unusual circular patch appears consult your vet. Meanwhile:
- Isolate the horse and all his equipment and tack; stop grooming; thoroughly wash your hands before contact with other people or horses.
- If the vet diagnoses ringworm, follow his instructions most carefully.
- Disinfect thoroughly the stables and equipment before use by another animal (page 264).

WARBLES
Warbles are caused by the maggot of the warble fly which in the course of a few days, bores a small hole in the skin and pops out.

Treatment

- Warbles are best left alone until the maggot has made its exit.
- It is important not to put pressure on the warble while it is developing as this can kill the maggot under the skin, leaving a permanent and troublesome swelling.
- You must *not* ride your horse if the warble is in the saddle area.
- Hot fomentations (page 272) will help to bring the maggot out. On no account squeeze it with your finger.
- Once the maggot has emerged, treat as a wound (pages 275–6) Stages 2 to 4.

Poor Condition

This is a most important subject, and worthy of a section of its own. The horse that fails to maintain condition in spite of every care is a source of worry to his owner and a matter of reproach from his fellow horsemen.

DEFINING CONDITION

Though horsemen employ many terms to describe condition carried by a horse, the more commonly used are:

Good Condition Body well covered, coat sleek and a general appearance of well-being.

Gross condition The horse is over-fat, particularly in the crest and loin regions.

Light condition Rather short of the ideal.

Poor condition Ribs in evidence and quarters short of muscle.

Debilitated Very short of flesh, coat dull and staring, general weakness and the horse unusable.

Emaciated Meaning 'all skin and bone'—extremely rarely seen in this country.

Other terms used are:

'Good Doer' A horse naturally maintains good condition without any special care and attention. Cobs are often good doers.

'Bad Doer' A horse who in spite of every care and attention never really looks well, an explanation for his condition being difficult to find.

'Dainty Feeder' A horse who is unusually fussy about what and how much he eats and tends not to clear up his feeds. In nearly all cases an explanation exists, but it may take weeks or months of painstaking study to arrive at the reason and to overcome it. Some horses, for example, will only feed at night; some only if a light is left on in the stable; some only provided that they are certain of not being molested by their neighbours at feeding times, and so on.

CAUSES OF POOR CONDITION

In order of importance, causes of poor condition are as follows:

Faulty Watering Arrangements

Nothing detracts from condition so seriously as this. An ample supply of fresh, clean water is essential to condition.

Some horses do not thrive on mains water due to chlorination. Stream or rain water drawn from a clean butt are best.

Faulty Feeding Arrangements

A horse will stand a great deal of work and still maintain

condition provided that the quality and the quantity of the forage offered is sufficient for the work performed and that the necessary time for feeding and a rest are allowed.

Possible Faults

- Insufficient or inferior quality forage is the most common cause of poor condition.

Stale, musty forage—both hay and concentrates—is not only unacceptable to the horse but is actually harmful, and its nutritional value is low.

The imperative need for adequate supplies of bulk in the feeding of a horse has been stressed elsewhere in this book, but it is not very easy to get a perfect sample of hay, as so often it is baled too soon after cutting and when in a damp state. Electric dryers and blowers to some extent offset this.

When buying hay, take the greatest care and if possible ask an expert for advice.

Half the problems of maintaining condition disappear when you can feed sweet, nutritious hay.

Among many other possible faults are:

- Irregular hours of feeding.
- Lack of rest periods in which to digest the feed.
- Bullying by a stable companion during feeding times.

Nutritional Deficiencies in Diet

Horses in poor condition may well be lacking certain minerals, vitamins or fibre, most of which are contained in old pastures with their mixture of weeds, herbs, reeds, grass, bushes, etc., which modern farming methods have largely eliminated.

Indications of a craving for minerals are:

- Licking soil.
- Gnawing wood or bark.
- Eating dung.

Treatment

- If the horse is at grass, the necessary herbs, etc., should be in your paddock unless it has had horses on it for too long, is horse-sick, or has only recently been sown.
- If the horse is in the stable, be sure he has a salt-lick.

In either case, if you still suspect mineral deficiency consult your vet.

Age

Age obviously affects the maintenance of condition.

The crisis in the life of a horse shows at about 15 years when his condition becomes more difficult to maintain, and he is up to less work.

Old horses suffer greatly from cold and require special care in winter time. Nevertheless, many horses pass this crisis without showing any deterioriation beyond a gradual slowing of pace. Such horses may well go on and lead a useful life up to 25 years of age or more, provided of course that they are not suffering from any physical disability or disease.

Horses over the age of 15 years are liable to deteriorate for a variety of reasons and therefore require extra care and attention. Ideally they should be checked annually by a vet both in their own interests and for the safety of the rider.

Unhealthy Teeth

Defects in a horse's teeth are often the cause of loss of condition, as they make his attempts to chew his food properly painful.

The grinding process carried out by the back (molar) teeth causes sharp edges to develop on them, which cut the tongue and cheek.

Broken or split teeth can cause much pain.

Precautions

Have your horse's teeth inspected once a year by your vet or some other experienced person.

Treatment

Tooth troubles are fairly easily put right. Sharp edges are easily removed by rasping, which your vet will do for you.

Worms

All horses harbour worms, and if these are not regularly dealt with they can cause serious and lasting internal damage. If present in excessive numbers they contribute to poor condition.

Many types of worm infest horses, the most common being:

Red Worm The most usual and the most dangerous. It makes its nest within the lining of the bowel, thereby destroying its function. It also inhabits the blood vessel walls and is sometimes the direct cause of the blocking of blood vessels, aneurisms and strangulation. In extreme cases it causes diarrhoea and emaciation. It is small and difficult to see with the naked eye.

White Worm Large and easy to detect, it causes few problems in adult horses.

Seat or Whip Worm Indicated by a soiling of the dock region.

Treatment

- Each year better and better worm drugs are discovered which are more lethal to the worms and less harmful to the horse. Seek expert advice as to which to use and at what intervals.
- The quantity and type(s) of worms can be ascertained by sending a small sample of droppings for testing. An agricultural chemist will do this for you. It will be unnecessary if you give your horse regular worm doses, which is much the best precaution.
- Horses should, if possible, be wormed in a stable. If it is not possible, droppings should be collected and disposed of to avoid contaminating the field.

Disease

Internal growths and consumption are often the cause of loss of condition in horses of 15 years or more.

Consumption in a horse is a disease of the belly and bones and not of the chest.

If your horse, whatever his age, tends to 'fade away' in spite of a ravenous appetite, you should take expert advice.

4

TRANSPORTING HORSES

By Horsebox or Trailer
By Air

By Horsebox or Trailer

Most horses travel perfectly happily in horseboxes or trailers, provided that they have been sensibly and sympathetically introduced to them and not subsequently frightened by inconsiderate driving.

Horses are sensitive to mood, so avoid showing signs of nervousness, haste or excitement.

Allow plenty of time for loading and unloading.

PREPARATIONS

It is most important to check periodically that the floorboards of the box or trailer are not showing signs of rotting, that the tyres are in good condition and that the brakes are sound.

Incorrect tyre pressures and faulty springs can cause towing problems and discomfort. Make sure that, unless provided with floor matting, the box or trailer is well bedded down with straw or shavings, and that all fitments are secure, well-oiled and easy to operate.

Always check your trailer coupling, ramps, side doors, legs and lights (including indicators).

Clothing

The number of rugs needed, if any, will depend on the weather. Remember that a horse generates a good deal of heat in a confined space, and many sweat with excitement. It is often best to use an open-weave type of sweat-rug, which should be secured at the front and be covered by a day rug with the front folded back and lying flat under the roller or surcingle. Alternatively, fasten the buckle at the front of the day rug. Rugs must be firmly secured with a surcingle or roller to prevent them slipping off and becoming entangled in the horse's feet, which may cause him to panic.

Always put on a tail bandage and/or guard to prevent damage to the tail. Never travel a horse in studs other than road studs.

On a Short Journey

When going to a meet or rally, etc, there is no reason why a horse should not travel partially or completely saddled up. If the horse is travelling with a bridle on, always put a headcollar over the bridle and do not tie him up by the reins which should be over his neck—the slack being taken up with a knot, or twisted round his neck, if necessary. If the reins are very long, they may be looped behind one or both stirrups.

On a Long Journey

It is advisable—and especially so with a young horse—to fit bandages, knee caps and, in the case of a doubtful traveller, hock boots as well, to prevent injury. You should bandage over gamgee, which should come right down over the coronet, to avoid damage from 'treads'.

Should the journey necessitate a period of eight hours or more continuously in the box or aircraft, it is as well to feed a bran mash the night before to guard against constipation. Thereafter feed as appropriate to the horse's condition, fitness and forthcoming work programme.

LOADING

Never ride a horse into a trailer or horsebox. Always lead him straight up the ramp.

- Look straight ahead yourself and don't look round at your horse.
- If the horse hangs back do not pull at him.
- Do not tie the horse up until you or an assistant has put up either the bar, the breeching straps or the ramp. If you do not do this, he may run back, fight against the rope, panic and do himself a serious injury: it is better for him to go loose.
- When transporting a single horse, put him on the off side of the trailer or box as he will travel more smoothly on the crown of the road.

With an Assistant

- *Either* come in on a circle from the side and lead the horse straight up the ramp. This is usually the most effective method. The assistant should then put up the ramp having first secured the bar or breeching straps, if fitted.
- *Or* position yourself and the horse not less than 4.5m (15ft) from the box or trailer and facing the ramp; then lead him straight up it.

Without an Assistant

- Lead the horse up the ramp and persuade him, if he will, to go in ahead of you. Secure the breeching bars or straps and either go up alongside him or round through the door at the front to tie him up.
- If he won't go in ahead of you, lead him in, pass the rope through the ring without securing it and return to the back, either alongside the horse or through the door at the front. Secure the breeching bars or straps and put up the ramp. If the horse persists in running back while you do this you must find someone to help you.

Horses Reluctant to be Loaded

There are many degrees of reluctance, but assuming that the horse has not previously had a fright, your success in coaxing him into a box or trailer depends largely on the confidence he has in you and the discipline you have instilled.

First decide why he is resisting and adjust your methods accordingly—Is it fright? Is it unwillingness to leave stable-companions? Is he just stubborn?

The following are some of the methods (there are many others) that may help and you should select the one that seems most likely to overcome the resistance of the horse in question.

- Encourage the horse with a feed or tit-bits.
- Either move the back of the partition to one side, thus making the entrance larger, or remove the partition altogether.
- Open the front of the trailer if it is a 'front unloader' and allow in as much light as possible. A shy horse will sometimes enter if he can see the way out.
- Load another horse first as a lead.
- Draw up alongside a wall, a hedge or another lorry, as this

makes the box or trailer look less forbidding than when it is parked out in the open. It also provides a 'wing'.

- Have an assistant walking behind the horse. This may be sufficient by itself, but the assistant should carry a lunge-line or whip, as a quick tap often makes up the horse's mind for him.
- If the ramp is too steep, lessen the angle by backing the box or trailer up against higher, level ground, and lower the ramp on to it. The end of the ramp must be firmly on level ground.
- Fit ropes to the back of the box or trailer and have an assistant on each side of the horse to hold them taut. This helps to keep a horse straight and often persuades him to go in. As the horse moves forward the assistants should change sides, keeping a safe distance from his heels with the ropes taut—if necessary applying pressure on his legs with them, above the hocks. If you have no assistant, attach a lunge-line or long rope to the off side of the headcollar, pass it round his quarters above his hocks and move back to his head on the near side, keeping the line taut with the left hand. Lead the horse forward in the normal way by the head-rope with the right hand; if he hangs back, encourage him to move forward by increasing the pressure on the lunge-line with your left hand.
- Have two assistants walking behind holding a short length of rope between them and if necessary applying pressure on his legs above the hocks.
- Dip a broom in water and flick the water and/or the broom on his quarters as he approaches the ramp.
- Blindfold the horse.

There are many other methods, but in the end it comes back to the confidence that the horse has in you, the way he has been transported previously and your own 'feel' and 'horse sense'.

Always be firm, but avoid using force which although sometimes effective the first time, creates quite the wrong atmosphere and seldom succeeds in the long run.

To Train a Horse to Load Willingly

When introducing a young or inexperienced horse, or one that has never been stabled, to a box or trailer, allow ample time for his own senses to dispel any doubts he may have, undisturbed by persuasion.

Open the front of the trailer and remove the partition. Let him, in his own time, follow another horse or the person leading, if possible straight through and out the other side; this process should be repeated several times. Finally, reward him with a feed or tit-bit.

To gain the horse's confidence it is helpful to feed him for a few days in the box or trailer without moving off.

UNLOADING

Front Unloading

- Untie the horse or horses before lifting the front bars in case they try to rush out. If more than one horse is travelling in the box or trailer, unload the horse nearest to the ramp first. Before unloading the second horse, move the front of the partition across to prevent him scraping himself against it or getting stuck.
- Lead the horse straight and steadily down the ramp.

Rear Unloading

- ***With an assistant*** Untie the horse before your assistant lowers the ramp and removes the bars or straps. The assistant should then stand on the ramp ready to keep the horse straight as he comes back by placing a hand on his quarters to prevent him stepping over the side of the ramp and scraping his hind legs.

  If the horse rushes back, on no account hang on to his head; all being well, your assistant will be able to catch hold of him, but if not, it is far better for him to go loose than to damage himself by rearing and hitting his head, or going over backwards.
- ***Without an assistant*** Untie the horse or horses before lowering the ramp and removing the bars or straps. If, having untied the rope, you leave it through the ring, many horses will think they are still tied up and remain still. You can then either catch the horse as he comes back or move up beside him to his head and back him out. Try to get him to come back steadily and straight, but do not hang on if he really rushes.

Alternatively, use a long rope or lunge-line to retain some control of him.

- ***Single horse in a double compartment*** When unloading from the rear of a trailer or box a horse will sometimes try to turn round. While there is room for a pony to turn, there is scarcely room for a horse to do so and he should, if possible, be prevented from attempting this. If you have no assistant, when you have untied him (leaving the rope through the ring) and having lowered the ramp, you must be very quick in moving up beside his head. To prevent the horse trying to turn round during the journey he may be tied to both sides of the compartment.

THE JOURNEY

Re-check the trailer coupling, ramps, side doors, legs and lights. As a horse is unable to see out, he cannot anticipate the movements of the vehicle, so drive at a steady pace—not more than 48–64km (30–40 miles) per hour—and as smoothly as possible, avoiding jerks when you stop and start, violent braking, and fast cornering.

Horses travel more contentedly if they have something to eat. Except when you are on the way to a meet, rally, competition or race, you should tie a haynet in front of each horse in the box or trailer.

Horses Travelling Together

Provided they get on well together and are not confirmed biters and kickers, most horses travel quite happily without a partition, which may be useful if one of the horses is a bad traveller and will not settle in a single compartment. However, there is a risk that the horses may tread on each other, which, if their shoes are fitted with road-studs, could be serious. For notes on travelling boots, see page 123.

Bad Travellers

Once a horse has been frightened, he will seldom regain his full confidence when travelling. Loss of confidence is usually the result of being loaded in a compartment that is too narrow; of being taken too fast round corners; or of sudden braking by the driver.

AFTER TRAVELLING

Always muck out your box or trailer, making sure that you clear the drain-holes, and push the dry straw up to the front of the box to allow the floorboards to dry, thus lessening the risk of their rotting.

Travelling by Air

Horses usually travel quite happily by air provided that they have a compartment that is wide enough to allow them to spread their legs and maintain their balance.

A capable attendant able to deal with panic or fright must accompany a horse. Attention should be paid to clothing. It is particularly important that suitable rugs, boots or bandages, a tail guard and a poll-pad are worn. The poll-pad is designed to protect the poll region from blows when passing through low doorways or under low beams, and to minimise the development of the injury known as poll evil.

Before undertaking a journey by air, seek the advice of someone with knowledge and experience of this form of travel.

Glossary

Advanced horse A horse that has reached a high standard of training.

Aids The means by which a rider—through his hands, legs, etc.—communicates with his horse.

Azoturia The muscles of the horse's back become tense, hard and painful.

Balance Distribution of weight between horse and rider.

Blistering Inducing severe inflammation of the skin to assist the healing process.

'Bone' A term used in conformation referring to the circumference of the bone below the knee. If the measurement is generous, the horse is said to have 'good bone' or 'plenty of bone'; if not generous, the horse is 'short of bone'.

Bounce fence Two fences with no non-jumping stride between them.

'Breaking out' Starting to sweat again after exercise, having previously cooled off.

Bridoon One of the two bits used with a double bridle. A bridoon rein is the rein that is attached to the bridoon bit.

Bringing up (Brought up) Bringing a horse, previously kept in a field either permanently or for a short period, to live in a stable.

Brushing Striking into the fetlock joint or coronet with the shoe of the opposite foot.

Bursal enlargements Soft visible swellings.

Cadence A horse going with rhythm and impulsion, producing 'elastic' steps.

Calkin A raised and squared thickening of the metal of the hind shoe at the outer edge of the heel, which gives increased grip. A similar effect is produced on the inside heel by thickening the metal into a wedge shape (wedge heel) which is less likely to cause brushing than a calkin.

Cast (a shoe) A horse is said to have cast a shoe when the shoe comes off by accident: i.e. it is not removed deliberately.

Cast (in box or stable) A horse lying down and unable to get up, usually as a result of rolling and getting jammed in a corner or getting his feet caught under the manger.

Cavesson headcollar A headcollar with a strengthened noseband, suitable for lungeing.

Cavesson noseband The standard type of plain leather noseband.

Centre line (of a manège, arena or school) The line between A and C.

Centre line (when approaching a jump) An imaginary line running along the ground straight through the centre of the fence or combination of fences.

Cheek (on a bit) The straight side-part of some bits.

Clench The part of a nail which during shoeing is left projecting from the wall of the hoof after the end of the nail has been twisted off. The metal is then bent over and hammered in to secure the shoe to the foot.

Cold hosing The use of a gentle stream of cold water to reduce inflammation.

Collection Shortened and raised strides in walk, trot or canter. See also p. 34.

Concussion The 'jar' caused to the feet and legs of a horse by working on hard ground. Also a medical term used to describe certain injuries to the brain.

Conformation The term used to describe a horse's physical characteristics.

Coronet A sensitive band, similar to the quick in a human nail, around the top of the wall of the hoof. (See endpaper, 'The Points of the Horse.')

Cradle A light frame fitted around a horse's neck to prevent him from biting or licking wounds, blisters, bandages, etc.

Crest The upper line of a horse's neck. (See endpaper, 'The Points of the Horse'.)

Croup The upper line of a horse's quarters. (See endpaper.)

Curb A thickening of the ligament of the hind leg just below the hock.

Dishing A faulty action caused by 'turned-in toes'. When going forward the horse throws his front foot, or feet, outwards instead of straight ahead.

Diving (at a fence) An incorrect, unbalanced jump. At the moment of take-off the horse's weight is on his forehand because his hocks are not actively engaged beneath him.

Dock The bone of the tail.

Double bridle A bridle with two bits and two sets of reins.
Drop noseband A noseband that is rather narrower than the cavesson noseband and is fitted below the bit.
'Dumped' toe This occurs when the wall of the hoof is rasped and rounded at the toe so that it fits the shoe.
Extension Refers to the action of the extended trot or canter.
Falling in This occurs when, in order to compensate for stiffness or loss of balance, a horse turning on a circle or round a corner moves his shoulder in and comes off the true circle.
'Fiddling' Interfering hands that continually change the contact both on the flat and during the approach to a fence.
Flat (jumping) The horse not lowering his head and neck and rounding his back: i.e. 'jumping like a deer'.
Frog The part of the hoof that, as it comes into contact with the ground, acts as a buffer to absorb the impact and prevents slipping.
Galls Sores and/or swellings.
Gamgee Cotton wool encased in gauze.
Gaskin The part of the hind leg above the hock. (See endpaper 'The Points of the Horse'.)
Going Depending on context this is used to describe the condition of the ground or the way in which a horse is behaving or performing. The 'correct way of going' refers to the correct movement and outline of a horse while being ridden.
Ground-line The base of a fence from which the horse and rider judge the take-off zone.
Hackamore A bitless bridle.
Hard mouth A permanently damaged mouth, caused by wearing down of the nerves.
Hobday's operation An operation to clear the obstruction in the larynx of a horse whose breathing (wind) is impaired.
Hollow back A back which is unduly dipped.
Hooked billet A form of mounting used to secure the rein to the bit. It consists of a small metal hook let into the leather about 6″ (15.24 cm) or 7″ (17.78 cm) from the end of the rein which also has a slit in it about 2″ (5.08 cm) from its end.
'Hot up' A horse which becomes unduly excited when ridden is said to hot up.
Independent seat A rider is considered to have an independent seat when his seat, body, legs and hands can all move independently.
Jabbing/jobbing the reins Jerking a horse in the mouth, either voluntarily or involuntarily.
Keepers Fixed loops, used to keep the ends of the straps of the bridle in place.
Lameness Unevenness of the horse's stride when moving (i.e. limping).
Left rein To be 'on the left rein' means moving to the left.
Leg-plate The lower of the two blades on a clipping machine that is coarser than normal. It is sometimes used on hunters to avoid cutting the hair too close to the skin.
Ligaments Strong, fibrous bands connecting bone to bone.
Loins The lower part of the back, just in front of the quarters. (See endpaper, 'The Points of the Horse'.)
Long and short sides Terms used to differentiate between the sides of a rectangular arena, manège or school.
Lumbar muscles The muscles of the loins. (See endpaper, 'The Points of the Horse'.)
Manège An enclosed (fenced in) area which is usually rectangular, but may be circular. It is used for exercising, taking a ride (mounted class), etc.
Moment of Suspension The moment when a horse has all four feet off the ground either over a fence or on the flat.
'Nappy' A 'nappy' horse is one who is stubborn, wilful, obstinate and unwilling to go in the direction required. Nappiness usually takes the form of refusing to move forwards, rearing (standing on the hind legs) or whipping round (turning round quickly and without warning).
Near and Off sides When facing to the front, the 'near' side is on the left and the 'off' side is on the right hand side of the horse.
New Zealand rug Waterproof canvas rug for use on a horse turned out in a field.
Novice horse An inexperienced horse.
Numnah A pad worn under the saddle to protect the horse's back.
Off side See Near and Off sides.
Over-face Asking a horse to jump an obstacle beyond his ability or stage of training.
Over-reach A wound which occurs when a horse strikes into the heel of a foreleg with the toe of a hind leg.
Pastern The portion of the leg between the fetlock and the hoof. (See endpaper, 'The Points of the Horse'.)

Pelham A type of bit. A pelham bridle is a bridle fitted with a pelham bit.
Pointing A horse standing with one foreleg stuck out markedly in front of the other. This usually indicates discomfort in the foot. The same action with a hind leg is normal and is referred to as 'resting a hind'.
Poll The top of the head between the ears. (See endpaper, 'The Points of the Horse'.)
Pritchel A sharp-pointed tool used by a farrier to make nail holes in a shoe while the iron is hot.
Puller A horse that pulls on the reins and is hard to stop.
Pulling (mane or tail) Removing hairs from the under-side of the mane and the sides of the tail to improve their appearance. (See also 'Puller', above).
Quartering Tidying up a horse in the stable.
Right rein To be 'on the right rein' means moving to the right.
Rolled toe When the toe of the hind shoe is drawn up and over the front of the hoof of a horse which drags its hind feet.
Roller A girth of leather or webbing with pads on either side of the withers to prevent pressure. It is fastened around the horse to keep the rug in place.
Runners Leather loops which slide up and down and are used to keep the straps of the bridle in place.
Running up in hand Showing off a horse, with him moving as freely as possible at the walk and trot.
'Run up light' A horse who has lost condition and weight and whose under-line runs sharply up to the stifle is said to have 'run up light'. When after strong work (such as a race or a day's hunting) only the underline is affected, the horse may be described as being 'tucked up'.
Run up a stirrup iron Slide the iron to the top of the leather.
Saddle tree The foundation upon which the saddle is built.
'Safed off' shoe A hind shoe where the outer ground edge is rounded, and the shoe then set back under the foot to minimise the risk of damage to a foreleg, i.e. over-reaching.
'Scalded' back Inflammation or, in severe cases, blistering of the back under the saddle, caused by heat and sweat.
Scope Athletic ability.
Set Fair Giving the horse a light brush over, removing droppings and tidying up bedding.
Short side (of manège) See Long and Short Sides.
Snaffle A type of bit. A snaffle bridle is a bridle fitted with a snaffle bit.
Speedy cutting Striking into the inner and lower side of the knee with the inside of the toe of the opposite foreleg.
Spring tree A saddle tree with a strip of metal let in at the waist to give more flexibility.
Stale To urinate.
'Staring' coat A coat standing up and looking dull instead of lying flat and looking glossy. It is usually a sign that the horse is cold or unwell, or both.
Stifle The joint at the top and to the front of the hind leg. (See endpaper, 'The Points of the Horse'.)
Straight shoulder This occurs when the line of the shoulder from the point to the withers (see endpaper, 'The Points of the Horse') is comparatively straight rather than sloping well back.
Strike-off The first step of the canter.
Striking into Scraping or knocking one leg with another. See also 'Over-reach' and 'Speedy-cutting'.
Surcingle A belt or girth of leather or webbing that is passed around the horse and fitted over the rug or saddle to help keep them in place.
Tendons Tough, fibrous cords, slightly elastic, that attach muscle to bone and give support to joints.
Tree See Saddle tree and Spring tree.
Tush In the male horse, a tooth behind the corner tooth on each side of the upper and lower jaws.
Turn out, to To put a horse out in a field (out to grass). It also refers to the process of producing a horse saddled-up and ready to be ridden.
Turnout, the The general appearance of horse and rider—grooming, trimming, saddlery, dress, etc.
'Underneath' a fence A horse is said to be 'underneath' a fence when the point of take-off is close to the fence.
United (true) canter The leading foreleg and leading hind leg appear to be on the same side.
Unmade mouth A mouth that is not obedient to the bit-aids because of lack of training.
Way of going See 'Going'.
'Wedge' heel See 'Calkin'.
Weaving A nervous habit that is incurable. It is sometimes caused by boredom. The horse rocks from side to side and may be seen lifting each foot

in turn as he does so. It is most frequently seen when a horse is looking over the door of his box and swinging his head from side to side. Other horses tend to acquire the habit from observation.

Well furnished In good condition, carrying good muscle and sufficient flesh.

Wind A horse's breathing or respiration when working.

Wings Extensions to the sides of a fence which are normally higher than the obstacle itself. They are aimed at discouraging a horse from running out. In the case of show jumps they also provide support for the fence.

Withers The parts of the horse that begin at the dip where the neck ends, rise slightly over the tops of the shoulder-blades and slope gradually downwards into the back. (See endpaper, 'The Points of the Horse'.)

Index

Pages on which relevant illustrations occur are shown in **bold** type

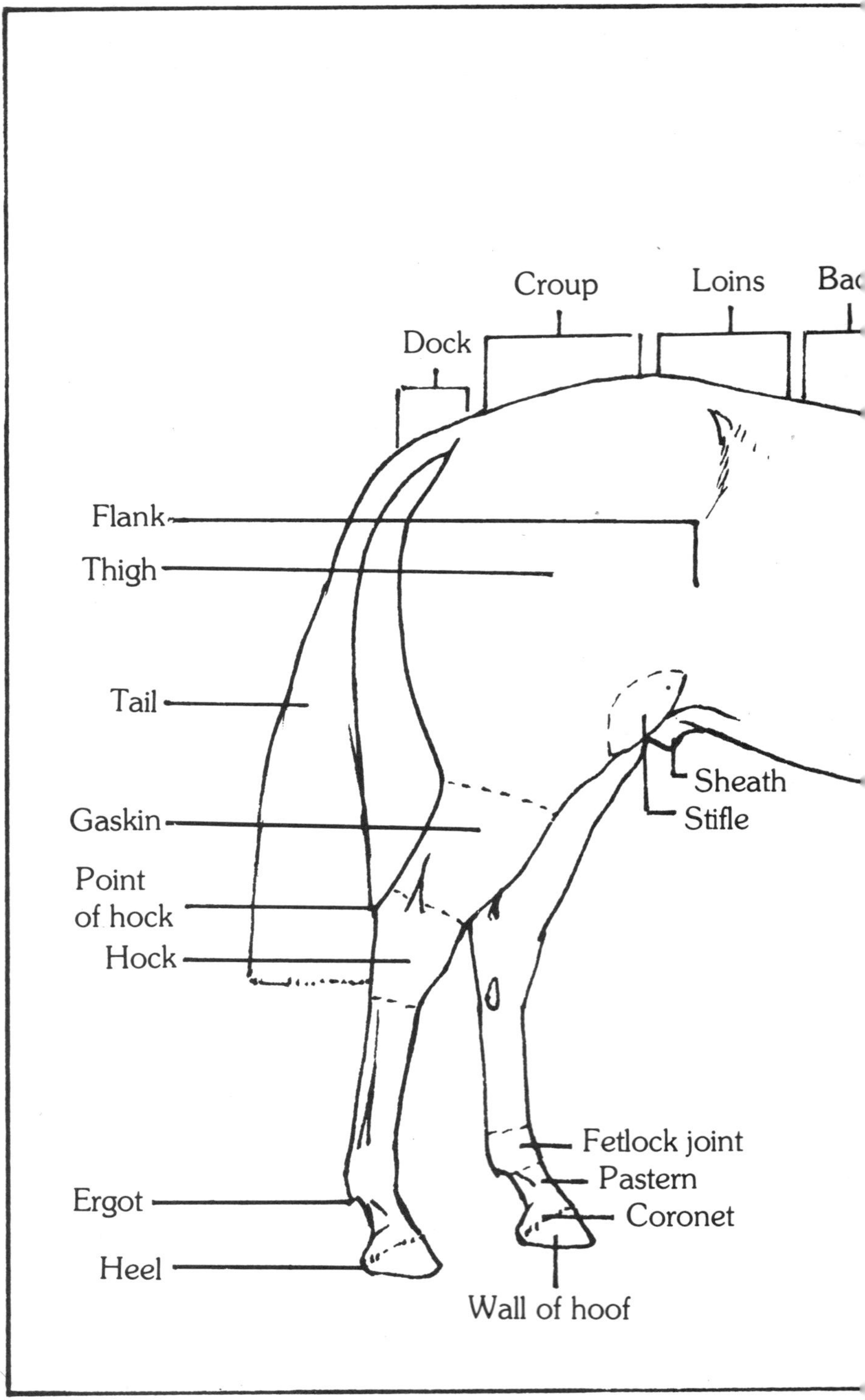

Croup
Loins
Ba
Dock
Flank
Thigh
Tail
Sheath
Stifle
Gaskin
Point
of hock
Hock
Fetlock joint
Pastern
Coronet
Ergot
Heel
Wall of hoof